THE HUB
PIONEERS OF NETWORK MUSIC

KEHRER

```
( \hubFilteredPulse,
    [\out,0,\freq,440.0,\cent
    \q,0.2,\atk,0.1,\rls,0.2
    \amp,1,\gate,1]

  )


DO: why don't I hear this sam
Samples = [ Buffer.read(
                s,
                "/Users/mark
                action: { "

            ) ];


partialSamples = [
                Buffer.read(
                    s,
```

CONTENTS

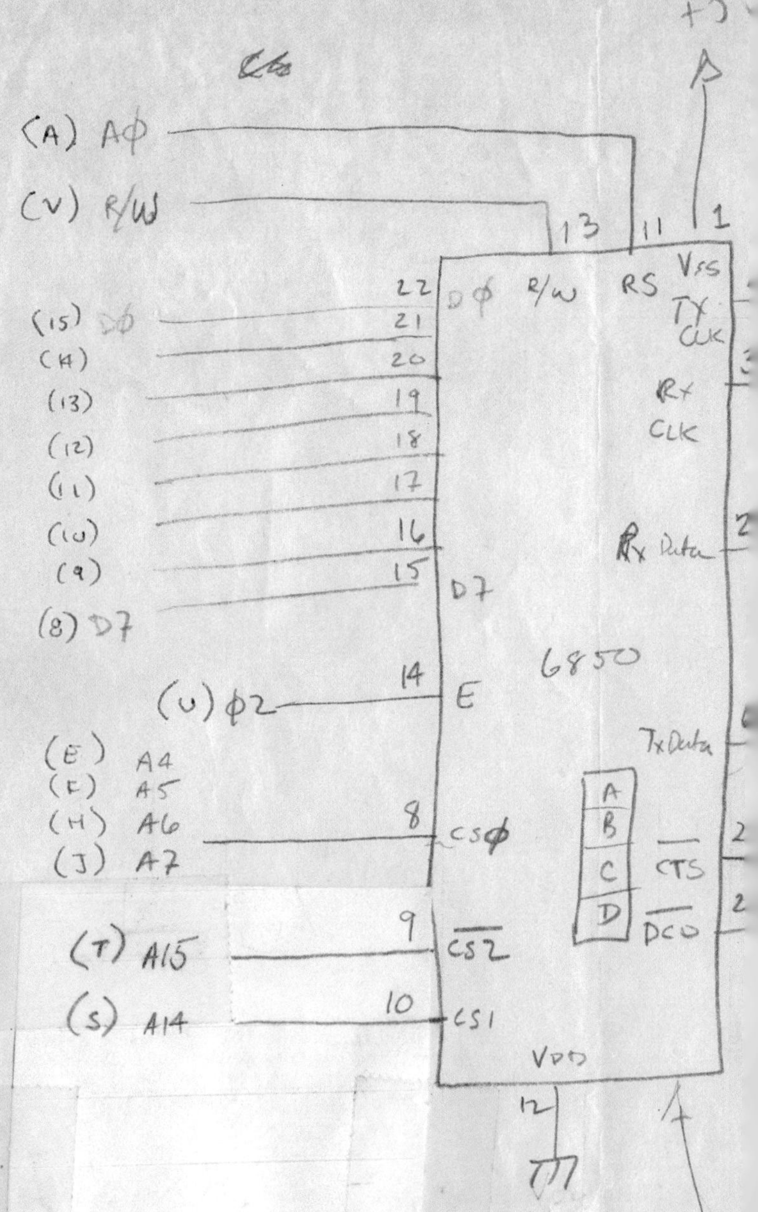

Xtal avail…

415

BUILD 4 of thes…
Multiple signal…
refer to signals…
on each chan…

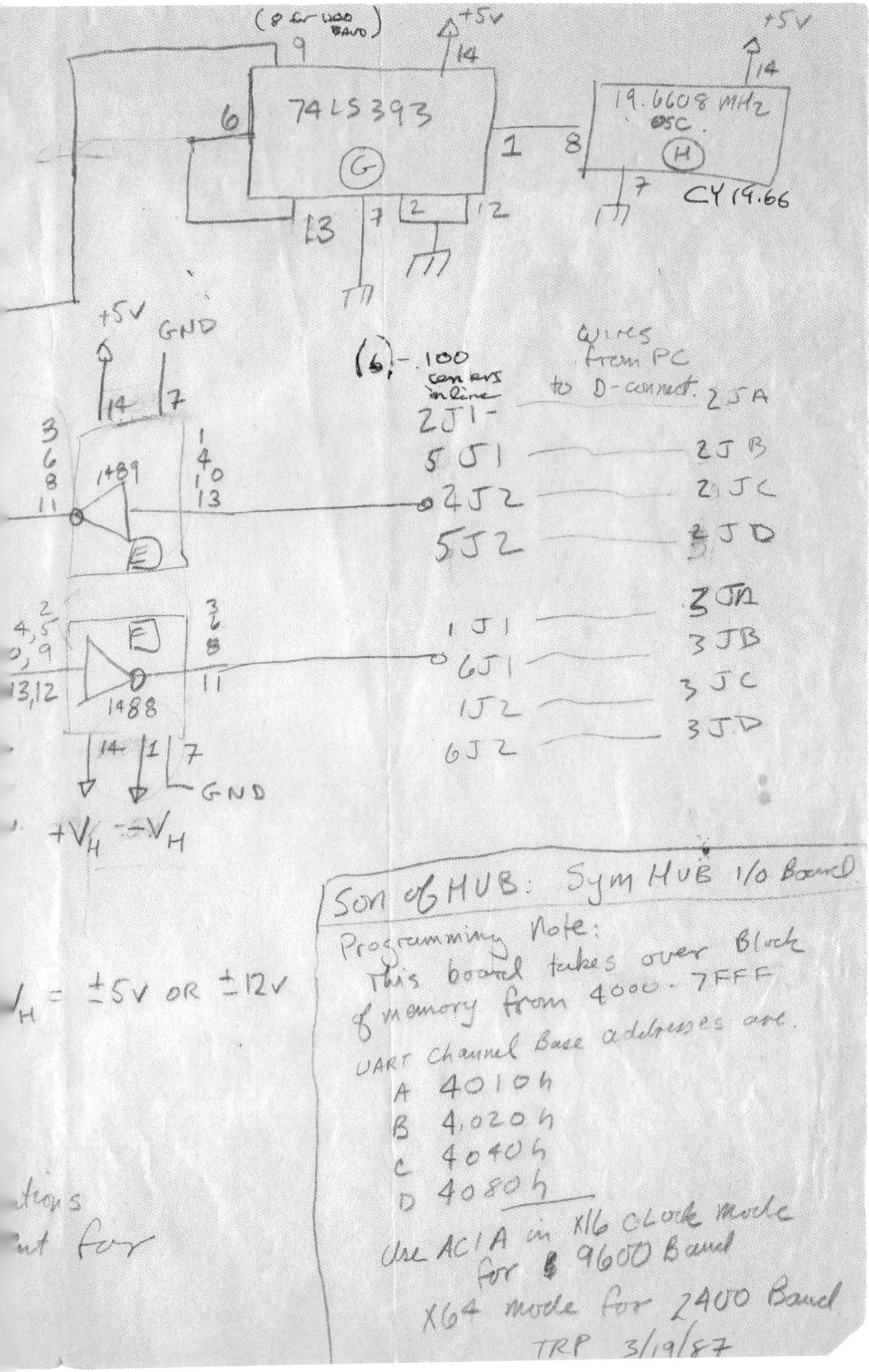

$V_H = \pm 5V$ OR $\pm 12V$

Son of HUB: SymHUB I/O Board

Programming Note:
This board takes over Block of memory from 4000-7FFF

UART Channel Base addresses are:

- A 4010h
- B 4020h
- C 4040h
- D 4080h

Use ACIA in X16 Clock mode for 9600 Baud

X64 mode for 2400 Baud

TRP 3/19/87

…tions

…nt for

OFFICIAL
Couch Potato
LODGE

CONNECTOR
POWER
UM6522A
8515
SYM
SYNERTEK SYSTEMS SYM
MODEL 1

CONNECTOR
Synertek Systems Corp.
U1 74LS138
U2 7404
U4 74LS00
LD 1
LD 2
SAV 1
SAV 2
RST
ON
DEBUG
OFF
LD P
GO
SAV P
REG
WP
MEM
S DBL
CR
SHIFT
CALC
C
DEP
D
EXEC
E
FILL
F
JUMP
8
VER
9
ASCII
A
B MOV
B
USR 4
4
USR 5
5
USR 6
6
USR 7
7
USR 0
0
USR 1
1
USR 2
2
USR 3
3
© 1978 SYNERTEK SYSTEMS CORP.
MADE IN U.S.A.

FOREWORD

In 1986, when personal-computer technology was still in its infancy and the so-called ARPANET project (developed in a military context and the precursor of today's World Wide Web) was winding down, American musicians John Bischoff, Chris Brown, Scot Gresham-Lancaster, Tim Perkis, Phil Stone, and Mark Trayle founded the Hub, one of the earliest network-music ensembles. In the same period, in 1989, the ZKM | Center for Art and Media was founded in Karlsruhe which, from then on, was also dedicated to networks. This is evident from its first major museum exhibition, curated and directed by the author of this text, which bears the name "net_condition."

The Internet and the new infosphere, consisting of enormous amounts of data, have led to a new way of thinking and a new way of practicing—network thinking and networked collective productions. Nowadays, social life is based on networking. Individuals, households, and entire cities are permanently connected via telecommunicative processes. It is therefore only logical that complex systems and collaborative processes between people and media also take place in the field of art. The Hub can justifiably claim to be the pioneers of network music.

Music is traditionally created in dialogue between a person and an instrument or between several people and their instruments. This way of creating music can arise from the fixed notation of the score, or it can take the form of free improvisation, as in jazz, in which the musicians react to each other. In both cases, musicians are connected both mentally and acoustically. A radical change in this line of tradition happened with the advent of the computer, when algorithms and networks became the basis of musical composition. Now the musicians have an additional, physical connection. Creative music becomes networked music. The Hub is one of those groups that shaped this new interactive and participatory way of creating music between a collective and a computer.

In fact, the Hub was named after the central microcomputer that not only links the musicians and their devices, but also has an influence over the indeterminate musical outcome. Humans and computers are in a permanent communicative process while the music evolves. Accordingly, the computer is an equal component of the collaborative and interconnected musical ensemble, which is why the Hub's interactive conceptual art can be described as an early artistic human-computer interaction.

Human-computer interaction strongly influences digital media art: This becomes evident in current discourses on artificial intelligence and generative art projects, which consider the computer as an integrative component of artistic creation. The ZKM is at the center of these negotiations and reflects this "being with the media" in groundbreaking exhibition formats and innovative artistic works.

The Hub recognized the potential of these collective modes of production as early as the late 1980s and they continue to set standards today with their compositions, which are created interactively in the moment of performance.

With this book, the ZKM publishes the first monograph on the work of the Hub and gathers the scores—equally historical documents and playing instructions—for the first time. Accordingly, the publication writes a history of network music and understands itself as an instruction manual for future projects of networked (sound) art.

Peter Weibel

EDITOR'S PREFACE

When the jury selected the Hub as the winner of the Giga-Hertz Award in 2018, they were particularly struck by the innovative force that has emerged from the group's work. The music of the Hub cannot be heard in isolation as a sound event. Simply listening to the music is almost not enough to recognize the potential hidden within it, to gauge the artistic dimension and to understand the richness of the ideas.

It is therefore not surprising that the Hub worked with techniques for which a set of terms was only developed decades later. Terminology is only necessary when certain aspects develop greater meaning. The working methods used by the Hub became trends with a time lag, whereas they were part of the group's innovative routine as early as 1986. Thus, the Hub can certainly be described as an avant-garde in electronic music. Tim Perkis, himself, described the music of the Hub as "a cybernetic and revolutionary cousin to jazz."[1] He meant, specifically, the fields of net music, hardware hacking, circuit bending, and live coding, for which the Hub provided the blueprints. Each of these aspects offers its own sound-aesthetic and media-technological universe of possibilities. Accordingly, the wealth of ideas inherent in the works of this group is potentiated.

Of course, these ideas of the Hub did not emerge in a vacuum. One has to think about the (historical) contexts and discourses in order to understand the sources of sound-art projects and their development. Here, the place

1 Tim Perkis and John Bischoff, "The League of Automatic Music Composers 1978–1983," August 2007, http://www.perkis.com/wpc/leagueCDnotes.pdf.

of action is technology-savvy California, the universities of the Bay Area and some of the American composers working there, such as John Cage, Henry Cowell, Lou Harrison, or Harry Partch; but especially the group from which the Hub emerged, the League of Automatic Music Composers, a formation that existed from 1978 to 1983. Unlike the League, however, the Hub still exists today and its members can represent their heritage accordingly. They can be the mouthpiece of the time; communicating issues, ideas, and solutions. They can represent the historical dimension and thus point to their importance for contemporary music.

This book is about this current legacy. For what the composers of the Hub have developed is as current and fresh at present as it was at the moment of its creation. More clearly, there are artistic levels in their oeuvre that cannot be comprehended by musicians of present and future generations by simply listening to the recordings. There is also a need to replay, understand, and translate the conceptual scores in order to experience the achievements of the Hub and to be able to transfer them into one's own practice. For example, Scot Gresham-Lancaster's piece *Stuck Note* has already been reinterpreted and performed by various network bands.[2] For this reason, it makes sense to document the concepts and improvisation instructions underlying the compositions and, thus, not only make them available for scholarly re-use. They can also stimulate lively music-making or function as a blueprint for new concepts of group compositions. In addition, the practice of appropriating and continuing the compositions enriches the reception of the works, which can be heard either on CD or online using the QR codes in this book. Furthermore, this publication includes extensive commentary by experts and companions of the Hub, as well as short essays by musicians and musicologists, which are intended to enable the phenomenon of the Hub to be classified and to provide an orientating context.

All this is done under the premise that the works of the Hub should remain alive in order to be an inspiration for young composers. They are intended to encourage improvisation in groups and motivate musicians to seek and exploit sonic conceptual possibilities outside of well-trodden

2 Wikipedia, "The Hub (band)," June 2, 2021, https://en.wikipedia.org/wiki/The_Hub_(band).

aesthetic paths. In addition, many of the concepts can be realized with new digital instruments and software environments. Software such as Sonic Pi, ChucK, or Super Collider for Live Coding, hardware such as the Raspberry Pi environments and laptops connected to hubs offer a much more comfortable technical foundation for the tasks of the working concepts presented here than those conceived and used by the Hub.

It remains a major concern to bring musicians and composers, both nascent and emergent, out of their current contextual and aesthetic stereotypes to reveal the diverse sonic imagination that the Hub has been working with since 1986. Like any music that can be reinterpreted, these works can be re-adapted to the sonic and technical possibilities of the present and, like the current interpretation of classic works by Johann Sebastian Bach on a modern grand piano, enriched with the possibilities of these instruments. Here, the hardware infrastructure in the form of mini-computers with open operating systems, new controllers, and networks with cable or wi-fi offer tremendous possibilities to reinterpret the Hub's ideas and, as the Hub has also demonstrated itself, to be inspired by concept and technology. Something that I personally took away from the Hub's music is that beyond the perfect form, the perfect sound, there is a music in which communication with others, playfulness in technology, and constant amazement are possible.

Finally, I would like to thank the Hub and all the contributors for their work, especially Cecilia Preiss for her patient efforts to keep things balanced and in context, as well as Gino Robair for bringing patiently all the texts to perfection.

Ludger Brümmer

DEDICATION: MARK TRAYLE (1955–2015)

Viewed in one light, the Hub is a group of inventors. All of us have designed and built our own systems, in both software and hardware—competence with a soldering iron was as necessary as the ability to code in the early days of the band. Even in this context, Mark Trayle's creativity stands apart. His hack of the Mattel Power Glove, as well as his transformation of the mundane act of swiping a credit card into a musical performance, are prime examples, but the Hub pieces he designed also had that same "new angle", inventive aspect. His specs yielded some of the most revelatory pieces, often from quite simple premises. Mark had a knack for finding the essence of a thing, then manipulating that essence in a way no one but he would ever conceive of. His pieces enhanced the synergy of our collective-but divergent voices in a unique way that often turned our always-separated timbral expressions into a comprehensible, unified sound.

Mark's music is a noise of variable focus where sudden condensations of tone provide structure and non-tones act as exciters and embellishers. His Hub pieces juxtapose the clarity of an idea with the messy assortment of group response. We can hear the singularity of concept throughout but always within a context of diverse actualization. He creates a formality that invites play, and we can hear the two qualities evolve in poised balance. He reminds us of an

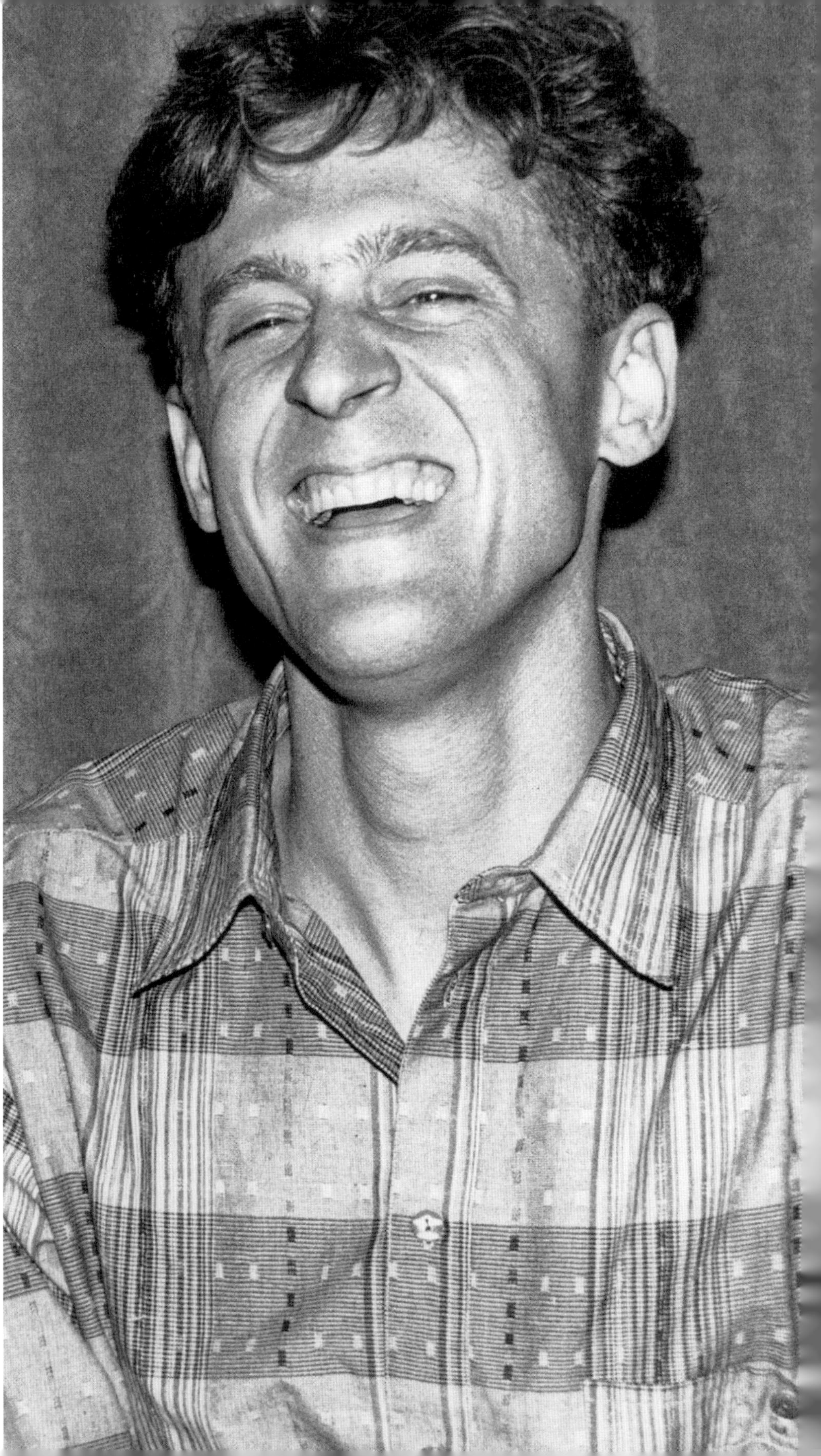

Mark Trayle, Photo: Barbara Golden.

eighteenth-century European composer in the transparency of his musical texture and his workman-like ease. He was a fan of Joseph Haydn, in fact.

Inventiveness alone doesn't explain the appeal of Mark's music. Another essential element is humor. Experimental music has its deadly earnest component; it can take itself very seriously, indeed. The best, however, can provoke joy and even laughter: a sonic Zen koan shooting like an arrow to the essence of the thing; Mark hit that bullseye again and again. As a bandmate, he kept us laughing and loose, and he made the hard work of rehearsal or the stress of touring seem easy and fun.

Joining the faculty of the California Institute of the Arts in 1996, Mark made important contributions to the development of the Experimental Sound Practices Program there as well as serving as co-chair of the Composition Program for a number of years. He also played often as an improvisor with personalized electronics as his instrument, using hacked electronic circuitry, physical controllers, and table-top electric guitar, all processed by home-brew software. His palette matured gradually from the sounds of monster trucks and video games to delicately balanced frameworks of resonated and distorted noise. He extended his work with the Hub in starting network bands in L.A. and collaborating internationally. Mark's strength as a teacher shows in the love that his students and peers have for him, and he brought the same spirit of adventure, inventiveness, and humor to teaching that he brought to our band.

The Hub by its nature is a form of digital glue that changes with every piece. Each of Mark's contributions added a unique voice to the realm of possibilities for our unpredictable interconnections. Mark is present every time we gather together, and we will never stop missing him. We dedicate this volume of scores to his memory.

Phil Stone, Davis, California, June 2021
(with input from John Bischoff, Chris Brown,
and Scot Gresham-Lancaster)

DEDICATION: JIM HORTON (1944–1998)

Without Jim Horton there would not have been the Hub. Jim Horton was a musician, composer, and polymath who saw early in the 1970s the potential of the newly available, inexpensive microcomputers to deeply change existing electronic music practice, pushing experimental music into the realm of artificial intelligence. With his introduction of the concept of the "Silicon Orchestra" to a Bay Area community of composers and musicians working with electronics, he catalyzed a way of thinking about the creative potential of machine learning that is still shocking and not completely realized today.

Horton was a creature of Berkeley, a sort of street hippie/punk/scholar, haunting the book stores and cafés of Telegraph Avenue, part of the glowing unofficial halo of bohemian eccentricity (and, in his case, untamable genius) that surrounds such university communities.

When I first met him in the late '70s, he was already physically a little beat and largely limited to his apartment near campus, a small studio jammed with eclectic stacks of books and smelling of fried hamburger (his appalling, sustaining diet) and Bugler tobacco. His seemingly limitless ability to craft wide-ranging tales and theories did not extend to revealing much about his own life story, and even after years of friendship, we still had only scattered images of his earlier life: childhood on a farm in Minnesota; there, using an old TV set and a barbed wire fence as an antenna for amateur radio astronomy; time at a seminary preparing for the priesthood; a dramatic auto accident with his "teeny-bopper" girlfriend plunging off a bridge into a river on one of their repeated trips bringing LSD from California back to hometown hippies; eventually moving to Berkeley and sleeping outdoors in the hills above the UC Berkeley campus for some years (which no doubt contributed to his premature aging and rheumatoid arthritis); and discovering experimental music by improvising all-night analog synthesizer concerts in industrial warehouses.

Dropping in on Jim at his apartment, one might encounter other astounding visitors—electronic musicians, certainly, or perhaps an improvising trombonist, a classics scholar or a science fiction writer; also local tinkerers, sculptors, celebrity food critics, wine experts, mathematicians, academic philosophers, and other eccentrics. One day, I met there a friend of his who shared a tale of the time Jim joined him at a party at a UC lab where they made hamburgers from an ancient frozen mammoth carcass liberated from the Siberian permafrost.

Conversation with Jim always resided somewhere on the continuum between dialogue and monologue. He would weave a theory/tale drawn from his wide reading about ancient tuning systems, Homeric poetry, the Montgolfier brothers, Incan hot-air balloons, mathematical logic, artificial neural networks (in the '70s!), evolutionary computer programming, post-serial music composition, the amazing punk band he heard on the radio last night at 3 a.m., CIA-funded abstract art galleries, and so on. Like his adored Herodotus, his tales always included the most vivid—no, lurid—details from every variant of a story, without worrying too much about consistency or plausibility. (Jim didn't make it into the twenty-first century, and at times I have wished I could hear his take on 9/11 or Trumpism.)

But really, Horton's ruling passion was his complete belief in the importance of the work we and he were all doing, applying what he saw as the most advanced thinking in computational logic and mathematical philosophy to the future of music. His vision of the "Silicon Orchestra" that led to the creation of the League of Automatic Music

Jim Horton preparing for a performance by the League of Automatic Music Composers at Fort Mason in San Francisco, 1981. Percussion instruments in the background were built by the ensemble Other Music who shared the concert.

Composers, the direct ancestor of the Hub, remained the unwavering guide for all his musical and theoretical work in the '70s, '80s, and '90s.

He always insisted that what we were all doing was in no way connected with entertainment but was research, scientifically sophisticated art on the cutting edge of human possibility. In some ways he saw the most important part of that work as the cultivation of advanced *listening*. He described himself as an "experimental listener." Commenting on his own music:

> *When the programs are running autonomously, slightly beyond my comprehension, playing music I probably wouldn't have thought of left to my own devices, I like to imagine they are precursors to uplifting, slightly alien musical A.I.'s of the twenty-first century. Oh, how I hope and wish that contemporary cyber culture will lead to a beautiful utopian compassionate world of Good!*

It is hard to emphasize how important Jim's perpetual optimism and excitement about new technically inspired culture was to us younger artists. The sense that one's work matters, that it is of unquestioned value, is a wonderful gift to receive as a young, perhaps not yet fully confident artist.

And though Jim was always in extreme poverty, this never seemed to discourage him, and I don't think he perceived it as a limitation at all. His dedication to his work was total: He would actually fast for several days if necessary to afford an expensive math textbook he felt he needed for his research. Happily packing for a gig, putting his rough-and-ready homemade music gear in a dresser drawer temporarily borrowed for the purpose, along with his single-board, '70s-era hobbyist computer (roughly equivalent in computing power to a current computer mouse), he felt he was doing important work pushing forward the state of the art. And he was right! No one else in the world was doing work like this at the time: It was essentially new in conception and did not depend on costly or sophisticated equipment.

As his health continued to deteriorate and trips to the hospital became more frequent, once near the end he experienced what he described as a vision after receiving an accidental medication overdose in the hospital:

> *Jesus on the cross. We zoom in and see that amidst Jesus's crown of thorns, there are myriad glittering lights, in fact an entire galaxy is rotating slowly around Jesus's head. Moving even closer we see that the points of light resolve to thousands (millions? billions?) of tiny gold Buddhas, each sitting in front of a glowing computer terminal. The computers are connected in a network spanning the entire universe, and are all engaged in stock market-like trading activity, constantly performing myriad tiny transactions, each designed—as is the whole network itself—to minimize suffering.*

It's worth saying again: Without Jim Horton, there would not have been the Hub. We all are very grateful to have had a chance to share some of our time here on Earth with such a unique and inspiring man and dedicate this book to his memory.

Tim Perkis, Berkeley, May 2021

ESSAYS

JON LEIDECKER
GOLO FÖLLMER
CARTER SCHOLZ

HUMANS IN THE LOOP
Jon Leidecker

Devices already have been perfected to record the impulses of the brain. In the music of the future, the composer will sit alone on the concert stage and merely THINK his idealized conception of his music. His brain waves will be picked up by mechanical equipment and channeled directly into the minds of his hearers, thus allowing no room for distortion of the original idea. Instead of recordings of actual music sound, recordings will carry the brain waves of the composer directly to the mind of the listener.
—Raymond Scott[1]

This music is electronic music based on Cybernetics, the science which sets forth principles of behavior which apply both to machines and people ... instead of writing musical notes to express our feelings, we design special electronic circuits. We feed instructions to these circuits. The instructions inform them how they should react and interact to the sensations they receive from other circuits, and how they should react to their own behavior, which they sample through feedback loops. This means that these circuits have an awareness of what other circuits are doing, what they themselves are doing, and what is expected of them by the designer—in this case, the composer.
—Louis and Bebe Barron (as quoted by Anaïs Nin)[2]

1 Raymond Scott, liner notes for *Manhattan Research, Inc.*, Basta 90782, 2000: 49.

2 Anaïs Nin, *The Diary of Anaïs Nin, Vol. 5: 1947–1955* (Boston: Mariner Books, 1975), 99.

Looking beyond the seemingly exhausted technology of written notation, some composers foresaw the imminent arrival of the medium of electronic music and described everything they expected these new tools to fulfill for them—namely, an even greater and unadulterated control over the materials of sound itself. In 1936, Edgard Varèse described what he expected from the future: score-reading machines, designed to "faithfully transmit the musical content to the listener."[3] And the following year, John Cage looked forward to music which could "reach the listener unadulterated by 'interpretation'... without the assistance of intermediary performers."[4] If sound recordings were a promising form of automation (albeit one which still required at least an original, capturable performance), perhaps they were only the first step towards an even more desirable form of music—an absolute work, that once conceived, would be obediently rendered in final form by the new tools, entirely uncompromised by any further human involvement.

By the 1940s, the engineering tools built to record and to measure sound had reached a sufficient level of sophistication to allow for the "discovery" of their creative uses. The dueling manifestos of France's GRM and Germany's WDR studios struggled to define the intrinsic nature of the new medium and the expanded worlds of control they offered to the composer. Pierre Schaeffer defined the new medium through its ability to capture and manipulate recorded sounds in ways which could liberate them for aesthetic use. For Herbert Eimert, the potential lay in the systematic measurement of all existing sonic phenomena, allowing the composer to synthesize any known or imaginable waveform, including sounds yet unheard. Yet, any apparent aesthetic divide between *musique concrète* and *elektronische Musik* quickly evaporated in the flood of compositions simply helping themselves to any and all technical approaches, as those studios were joined by an increasing number of other studios, each with their own distinctive approach to instrument design. Indeed, while the manifestos declare the will of the composer as paramount, the real key to understanding the music made during these early stages of development is to list the works not by the composer that created them, but by the studio in which they were made.

3 John Cage, "The Future Of Music: Credo," in *Audio Culture: Readings in Modern Music*, eds. Christoph Cox and Daniel Warner (New York: Continuum, 2006), 25–28.

4 Edgard Varèse and Chou Wen-chung, "The Liberation of Sound," *Perspectives of New Music*, Vol. 5, No. 1 (Autumn–Winter, 1966): 11–19.

In hindsight, it is the less-circulated, early '50s writings of Louis and Bebe Barron that grasp the real creative advantages and that touch on so many of the concepts we take for granted: generative music; self-playing instruments; ceaseless musical forms which one guides and interacts with rather than author or fully control. Their visits to the early Macy Conferences exposed them to the emerging field of cybernetics, suggesting an entirely different paradigm for musical control from the outset.

Norbert Wiener's insight that self-regulating feedback systems, capable of adjusting their behavior through the self-observation of their outputs, were already exhibiting characteristics easily definable as forms of intelligence. Louis Barron, experimenting with electronics and audio, went home from Wiener's lectures and routed the output of electrical circuits back into their inputs, driving them into audible self-oscillation. Adding variable capacitors and resistors to the circuits in ways that allowed them to interact with the sounds without controlling them outright, the Barrons discovered sounds that unmistakably resembled insect voices and birdsong—an immediate musical proof of concept for Wiener's theories. And whereas negative feedback was key to the design of simple self-regulating servo-mechanisms (such as thermostats and fuel gauges), perhaps there were creative uses to be found in positive feedback, where energy reamplifies itself as it loops, over and over, until it reaches, and thusly illustrates, the physical limits of a system (such as a vacuum tube or amplifier). The Barrons' experiments with feedback made a habit of pushing their equipment to the point of collapse to produce fascinating and evolving sounds with a greater expressive complexity than any of the laboriously edited, yet still largely rudimentary music of their contemporaries, leading to their earliest compositions, *Heavenly Menagerie* and *For an Electronic Nervous System* (1950–1953).

In the early 1950s, John Cage realized his early electronic works for tape at the Barrons's studio in Greenwich Village, regularly accompanied by his colleagues who assisted with the long hours of tape editing. One can only imagine the conversations that occurred in a studio dedicated to the design of autonomous musical instruments during the years Cage closed in on his formal practice of indeterminacy. It is likely they were played the Barrons's early compositions. And proof of a particularly critical collaboration from this time is found in the credits for Ian Hugo's 1954 film *Jazz of Lights*, where David Tudor's name is given as an assistant to the Barrons. Unlike their better known, highly composed cues for the 1956 Hollywood film *Forbidden Planet*, their score for Hugo deploys raucous outbursts of raw circuit feedback in a shockingly modern manner, anticipating Tudor's early '70s work with circuit feedback, as well as drawing a straight line to the piezo-and-microphone feedback systems that Cage and Tudor would soon make central in their works of "live electronic music" for the concert hall: *Cartridge Music* and *Variations II*.

Robert Ashley has called feedback "the only sound that is intrinsic to electronic music"[5] The supposedly transparent medium of reproduction is flooded, overdriven, until a native voice is revealed, those artifacts which could not have existed before circuits, amplifiers, microphones, and recordings.

In the wake of Tudor's realization of Cage's *Variations II*, we see a quick succession of creative works in which elemental principles of gain staging are taken to extremes that nearly have to be experienced as metaphors if they were not also on the edge of being physically dangerous, such as Ashley's *The Wolfman*, David Behrman's *Wave Train*, and Max Neuhaus's *Fontana Mix-Feed*. Extended tape delays become another form of feedback by incorporating recorded media into real-time improvisations, as in the works of Pauline Oliveros and Terry Riley (who reapplied his work with feedback delay into notation that helped give rise to the development of Minimalism), and the hybrid of tape delay and acoustic feedback demonstrated within Alvin Lucier's *I Am Sitting in a Room*.

Wiener's influence reappears in Gordon Mumma's name for his cybersonic circuits, capable of determining, and then storing, the most resonant frequencies of a room, then sending that room into overdrive by tuning electronic processing precisely to those frequencies. And in France, Éliane Radigue's interest in the endlessly sustained pitches intrinsic to sound generation with microphone feedback seemed

5 Thom Holms, *Electronic and Experimental Music* 2nd ed. (New York: Routledge, 2002), 27.

to naturally lead her to extended, ceaseless meditative forms in her works for tape. One critical aspect shared by all of these "compositions" is that there is no predictable sonic outcome. Rather than a fixed sequence of sonic events, the performer's actions serve to influence and illustrate larger complex systems, rather than claim to control them outright.

The introduction of commercial modular synthesizers in the mid-1960's brought a new level of public visibility to the practice of electronic music. Robert Moog's commercial success and decades-long influence over subsequent commercial synthesizer design is easily explained: He built ever-simpler, ever-more-limited interfaces that allowed a user to quickly "dial in" a sound with certainty. Don Buchla, on the other hand, intentionally cultivated coincidences, offering designs such as the Source of Uncertainty and the Multiple Arbitrary Function Generator, recognizing automated random-number generation as a powerful electronic tool, and allowing for a certain amount of user control over the desired amount of chaos.

These design goals were hardly limited to the United States. Gottfried Michael Koenig and Jaap Vink configured unpredictable modular systems capable of producing effectively endless performances in real time; many of their hyper-detailed works for tape are live captures. And in the Brussels basement of composer Leo Kupper lies his GAME synthesizer, designed in the 1960s; one particularly powerful module can be set to fire off a trigger at random intervals, with a maximum threshold of up to one week. As Kupper puts it: "This means that when performing on this machine, you would require as much as one week before you could perceive the structure of the music."[6] This synthesizer, almost undetectable beneath a forest of patch cables, appears in the packaging for German composer Roland Kayn's *Simultan*, along with this definition:

> *Cybernetic music, whether vocal, instrumental, or electronic, is above all that which is no longer 'composed' and rehearsed... the composer... completely loses his original function. He now merely decides whether to intervene, guide, and regulate, or whether he is prepared to accept that which emerges as an auto-generative process.*[7]

6 Leo Kupper, personal interview with author, Brussels, February 11, 2020.

7 Roland Kayn, liner notes for *Simultan*, Colosseum Colos SM 1473, 1977.

From the 1950s to the 1970s, the process of working with digital computers seemed bound to the compositional model by simple virtue of requiring extended amounts of rendering time to produce a score, let alone an audible waveform. Infinite variety was conceptually possible, though not in real time. Statistical and stochastic generation opened possibilities for new forms in the work of Lejaren Hiller, Iannis Xenakis, Max Mathews and John Pierce, James Tenney, Pierre Barbaud, and Sister Harriet Padberg. When personal computers and simple digital-audio processors started becoming available in the mid-1970s, the principles of feedback made way for real-time data streams.

Consider the algorithms capable of simple pitch tracking, which analyze an audio waveform and extract a monophonic melody from it to create a control signal capable of driving a distinct, synthesized voice singing along with it. Up until that point in history, hearing a tune sung by more than one voice at once implied preparation, rehearsal, tradition. Technology that is now capable of reflexively singing in unison, in under 30 milliseconds, liquidates our ability to distinguish composition apart from improvisation. And that is just the very surface of the kind of complexity these tools are capable of extracting and responding to in real-time performance.

That kind of complexity is certainly less meaningful when left to its own devices; an audience can only truly perceive the nature of the machine's listening if they can also see and hear a human performing, in real time, with a role to play as an intrinsic part of the larger network. Music that leaves room for a human in the loop becomes a way for us to understand and to model the evolving nature of the relationship between ourselves and our technologies. As Joseph Weizenbaum was trying to tell us with his computer program ELIZA, the question at hand is not what work machines are capable of automating: They're already capable of too much. The real question is: What is the work that should not be ceded to machines without doing damage to ourselves?

Over the last five years, advances in deep learning have led to networks capable of rendering "original" works of music in the styles of known artists. Unlike the work of Lejaren

Leo Kupper standing next to his GAME (Générateur Automatique de Musique Électronique) modular system in Brussels, Belgium, February 11, 2020. Photo: Jon Leidecker.

Hiller or David Cope, who manually translated baseline aspects of musical theory and compositional styles into code, a deep-learning project such as OpenAI Jukebox starts from scratch, reverse engineering its own understanding of the underlying rules of music strictly by training itself upon musical recordings. There are clear financial incentives guiding this kind of corporate work; an endless pipeline of new content in the style of musicians that audiences are already known to enjoy, regardless of whether those musicians happen to be alive. And there is certainly much to be learned by listening to these recombinant experiments hearing the kinds of patterns only computers can detect. Once trained on a data set, and given enough processing power, newly rendered waveforms can be produced faster than the time it takes to listen to them, creating recordings that seem to challenge our definitions of live music. One project by the Boston-based programming team Dadabots, entitled "Relentless Doppelgänger," has been continuously live-streaming algorithmic death metal since September 3rd, 2019. One of their slogans is "eliminating humans from music."[8]

One could easily make the point that humans are still in the loop, even if only as listeners, when regarding such works of generative music. A lot can be revealed about the world we live in by regarding the bizarre productions of neural networks that have ravenously ingested the history of our recorded music and found certain baseline structures within it that slide between the recognizable and the utterly alien. Such renders contain those kinds of fascinating mistakes that it would never even occur to humans to make, once again beautifully revealing our various assumptions about what music is.

Experiments in music are less dangerous than the experiments being made with autonomous weapons, which (for now) still legally require humans in the loop to make the final kill decision, but of course could be far more effectively fatal without them.[9] Later in his life, after several decades of creating unstable feedback circuits designed to fail in spectacularly interesting ways, Louis Barron's feelings of ethical responsibility to his instruments intensified: What precisely is being modeled by a music which uses the expressive "death screams" produced by his equipment as it self-destructed, occasionally bursting into flames while doing so?[10]

Rolling back to the beginning of the medium, there is a line leading through the work which recognized that the new tools were already more than neutral servo-mechanisms, and that the music made with them always expressed more than the personal will of the individual composer working with them: the resulting sounds are a model of the complex systems that created them. Or, as David Tudor put it when asked what he hoped an audience would take away from one of his concerts: "I would hope that it would be obvious to the listener that this was music made by humans."[11]

8 Singing and AI channel, "Dadabots: Neural Synthesis – Eliminating Humans from Music," YouTube, June 1, 2021, http://youtube.com/watch?v=Y2VWyk2suWw.

9 Will Knight, "The Pentagon Inches Toward Letting AI Control Weapons," *Wired*, May 10, 2021, https://www.wired.com/story/pentagon-inches-toward-letting-ai-control-weapons.

10 Phone conversation with David Barron, May 31, 2018.

11 This quote by Tudor is taken from personal transcriptions of interviews since lost and quite possibly apocryphal.

EPISTEMIC MUSIC

Golo Föllmer

The Hub is a computer network band. Computers, network, and band members each have specific roles in the system they form together. Individual algorithms on each of the computers are interconnected and compose the music collectively. The sound qualities emerging from the synthesizing modules are designed and regulated by the human members individually. What gives each piece of music of the Hub its conceptual framework and characteristic appearance is the type of network connection and communication instantiated in parameter connections and software code used by the machines and humans involved. Hub member Tim Perkis points out where this kind of connection between human, electronic, and digital agents necessarily leads to: "It used to be common to discuss the computer as an extension of the human nervous system; today, it makes more sense to think of the computer network as an extension of society."[1] Man-machine communication and its relation to communication in social systems play crucial roles in the music of the Hub.

A hub is a small electronic device that connects computers linked by cable to form a network. In 1986, a group of six musicians named themselves after this inconspicuous piece of technology. At that time, only a few people were aware of the existence of the Internet, and not even they could foresee the importance it would one day acquire. The form that the hub took in the work of the Hub changed over time. For the first CD, a dedicated computer was still

1 Tim Perkis, liner notes for *Wreckin'Ball*, Artifact Recordings ART 1008, 1994.

used as a hub. It not only provided the technical connection between the musicians' computers, but also evaluated data and intervened in communication. By the second CD they were using an Opcode Studio 5 MIDI interface as the hub, not because it offered more functionality, but precisely because of the limitations of the MIDI standard. Rule systems, their limitations, and interventions into them are topoi that appear central in their music.

This article proposes to interpret the musical activity of the Hub as an aesthetic practice performed with the aim of gaining knowledge about the society it is contained in. We can approach this idea using a concept laid out by anthropologist Steven Feld. After having researched the music and sound world of the Bosavi, an indigenous group in Papua New Guinea for many years, Feld coined the term *acoustemology*, joined from the words *acoustic* and *epistemology*, as an expression of his conviction that knowledge is not transmitted in chunks of essence through perceptive acts (perceiving for instance acoustic phenomena, including sound, noise, and music).[2] Instead, Feld applies John Dewey's relational ontology, assuming that "substantial existence is never effective prior to being-in-relation."[3] While essentialist perspectives assume that the world consists of entities which are composed of original substances for which we only need to apply the appropriate method of perception (or measurement) in order to fully grasp them, relational ontology understands listening and sound production as "knowledge in action." Acquiring knowledge thus requires engagement in the "relationality [...] of knowledge production in [...] contextual and experiential knowledge."[4] Being is thus never independent of being-in-relation, and knowledge emerges from relational interaction.

> *Acoustemology puts first stories of listening and tuning in through relational practices of listening and sound production and their reflexive feedback productions. [...] In stories of making sound, acoustemology hears a heterogeneous chance-bound narrative; in it, it hears stories about forms of coexistence, stories in which sound grounds a difference—radical or otherwise.*[5]

Steven Feld concluded that the life of the Bosavi is built on listening to all species. He interprets Bosavi chants as

2 Steven Feld, "Akustemologie," in *Handbuch Sound*, eds. Daniel Morat and Hansjakob Ziemer (Stuttgart: JB Metzler, 2018), 2–7. (Translation by the author.)

3 Ibid.

4 Ibid.

5 Ibid.

apparatuses for the group's coexistence. When the Hub's computers do their chant, whose coexistence may be at stake and which stories do they tell?

Let's start at the beginning, or even earlier, since the idea of the Hub started before its own beginning. What the Hub did with the first regular desktop computers, the League of Automatic Music Composers had previously done with home-soldered circuit boards and rudimentary precursors to the PC. Visionary Jim Horton infected Rich Gold and later Hub co-founders Tim Perkis and John Bischoff with his fascination for a "Silicon Orchestra" that would take on a life of its own and connect with humans. For the League of Automatic Music Composers, tables full of self-built and finickily wired electronics were not their instrument, but the artificial, non-human part of the "band." Each of the League's pieces was characterized by a special hardware and software configuration, which implied elements of a compositional rulebook inherited from the genetics of the applied technological structure and a consciously chosen compositional structure. The foundation of their setup was a KIM-1, the first affordable micro-computer.

In 1983 the League disbanded. Three years later, League members John Bischoff and Tim Perkis formed the successor group, the Hub, with Chris Brown, Scot Gresham-Lancaster, Phil Stone, and Mark Trayle. In the spring of 1987, the first trans-local concert took place in New York between the Clocktower and the Experimental Intermedia Foundation: There were three musicians at each of the two venues, connected via modem. While the Hub experimented with a number of dislocated setups, local network performances with all players at one table facing each other prevailed and proved most appropriate for the purposes intended by the Hub.

Various impulses of American experimental music can be found in the work of the Hub. With the self-image of the composer as someone who builds his own instruments, they connect to the music of Harry Partch and John Cage. With the understanding of electronics not only as a tool but as a musical actor that brings irritation and unsought finds, they connect to David Tudor and Gordon Mumma. Musical structures based on collaboration and group processes are linked to anti-consumerist counterculture that developed in the San Francisco Bay Area after the 1960s. Throughout time, the pieces and configurations have reflected the evolution of technologies, from board computers to the first desktop computers to laptops; from analog sound sources to MIDI instruments to settling on a diversity of software synthesis methods, while still, today, sometimes engaging elements of sound sources and interconnection techniques from earlier times.

For John Bischoff, the image of band members distributed in different rooms appears as the appropriate metaphor for the type and depth of interconnection between players of the Hub's precursor, the League of Automatic Music Composers. Even though all performances of the League occurred with all members at one table (or a bunch of tables), momentary musical arrangements seem to have felt like the performers were separated by walls, unable to see each other. Bischoff interprets the group's efforts as experiments in "styles of music interconnection."[6] That interconnection is a blind one, challenged by a much narrower communication channel than available to musicians showing each other their physical handling of a violin, clarinet, or voice. Bischoff's central question is: Which kind of "subtle interplay between unity and diversity"[7] results from which kind of interconnection? Interconnections are formed by decisions about which types of information to exchange between connected agents and what to do with them. Since usually there are many more channels of information excluded than included, interconnections are basically restrictions. Igor Stravinsky describes the concept of restriction as a necessary condition out of which compositional actions can emerge:

> *My freedom therefore consists in moving within that narrow framework which I have drawn for myself for each of my projects. I go even further: my freedom becomes all the greater and more comprehensive the narrower I mark out my field of action and the more obstacles I erect all around. He who robs me of a resistance robs me of a power. The more constraint one imposes on oneself, the more one frees oneself from the chains that shackle the mind.*[8]

6 John Bischoff, "The League of Automatic Music Composers," *Neue Zeitschrift für Musik*, 5/165 (2004): 24.

7 Ibid.

8 Igor Strawinsky, *Musikalische Poetik* (Mainz: Schott, 1949), 41–42. (Translation by the author.)

9 The Hub, liner notes from *Wreckin' Ball*, Artifact Recordings ART 1008, 1994.

10 Ibid.

11 Ibid.

12 Perkis 2004, n.p.

13 Johan Huizinga, *Homo Ludens: Vom Ursprung der Kultur im Spiel* (Hamburg: Rowohlt Taschenbuch Verlag, 1987), 146. (Translation by the author.)

14 Axel Volmar and Alexandra Supper, "Sonifikation" in *Handbuch Sound*, eds. Daniel Morat and Hansjakob Ziemer (Stuttgart: JB Metzler, 2018), 75.

The importance of this approach in the Hub's music is apparent in many of the piece descriptions. *Waxlips II*, for instance, is applying the same rule set as *Waxlips* "with an additional restriction of the transformation rules used by the players."[9] Perkis also emphasizes this when pointing out two general properties of their music. Opposite to most practices of music composition and production done today, their music is not based on "attempting to gain more control over every aspect of the music, [but to] seek more surprise through the lively and unpredictable response of these systems."[10] Likewise, the imperfect performer is not to be overcome, but his social component is to be emphasized and brought to the fore.[11] The piece *Hub Renga*, for example, takes up the traditional Japanese poetic form of *renga* where different poets contribute lines to a collectively built poem. During the aforementioned 1989 live performance, the composers' algorithms reacted to text streams coming from writers in a poetry conference on the WELL.

> *It's hard to describe the peculiar quality of community feeling that arose in this project: there was something authentically new about people being able to interact socially, while at the same time doing the traditionally more private work of writing and composing music.*[12]

By having the Hub base each piece on different rules, and thus foreground features of those rules, the music bears essential traits of the game as described by Johan Huizinga:

> *[The game] is an action that proceeds within certain limits of time, space, and meaning, in a visible order, according to rules voluntarily adopted, outside the sphere of material utility or necessity. The mood of the game is rapture and enthusiasm, and either a sacred or a merely festive one, depending on whether the game is consecration or amusement. The action is accompanied by feelings of exaltation and excitement, and carries with it merriment and relaxation.*[13]

Chris Brown sees the purpose of the Hub's networked musical activity as sonifying complex interactions between players. Sonification refers to processes of presenting data in the form of sound events for the purpose of conveying information via the auditory system.[14] Each band member's computer program plays its musical part independently and is regulated by its player on the one hand and by the communication processes between the computer hardware and algorithmic systems on the other. Brown's curiosity is directed toward obtaining a sonic image of what principles or dynamics of communication emerge when specific communicative processes interact with structuring processes. Computers started to become part of daily lived realities in the 1970s and 1980s, and networks started to frame interactions with computer processes on a larger scale in the 1990s. Brown's curiosity is about the state and evolution of this technology at any given time: "We sought to sonify, and listen to, the technology that was emerging around us."[15] In Bischoff's piece *Perry Mason in East Germany*, each musician is sending three music parameters into the hub and is obliged to use three parameters of any other musician at any time in his own sound producing algorithm. This can be interpreted as a musical application of protocols of machine communication that organize the Internet, but also of protocols that organize society. The music the Hub draws from this metaphor is not presented as a construction to be pondered, but as a process to be witnessed as it evolves.

According to Christopher Small's concept of *musicking*:

> *(t)here is no such thing as music. Music is not a thing but an activity, something that people do. The apparent thing music is a figment, an abstraction of the action, whose reality vanishes as soon as we examine it at all closely.*[16]

This musical practice is a socially responsible activity, a way of trying out alternative ways of structuring communication and relation as it is described in Jacques Attali's book *Noise*. For the French economist and philosopher Attali, music has been part of growth, the generation of capital and the staging of the spectacle ever since society's transition into a commodity exchange system. From that moment on, music has been an attribute of religious and political power and marks order, but also announces subversion.

> *Music is more than an object of study: it is a way of perceiving the world. A tool of understanding. Today, no theorizing accomplished through language or mathematics can suffice any longer; it is incapable of accounting for what is essential in time [...] It is thus necessary*

15 Chris Brown, "Wieso Netzwerkmusik?", *Neue Zeitschrift für Musik*, 5/165 (2004): 25.

16 Christopher Small, *Musicking: The Meanings of Performing and Listening* (Middletown: Wesleyan University Press, 1998), 2.

to imagine radically new theoretical forms, in order to speak to new realities. Music, the organization of noise, is one such form. It reflects the manufacture of society; it constitutes the audible waveband of the vibrations and signs that make up society. An instrument of understanding, it prompts us to decipher a sound form of knowledge.[17]

Imogine yee as tha rider could ontervoone in the chooce uf voowels used on me linguage. If that was the case, expression in the previous sentence would be controlled not entirely by myself, but in parts by the reader being able to change vowels following his or her own will or some rule. If we applied such kind of rule set to our verbal interaction, our communication would be close to how the Hub set up some of their pieces. While in our understanding of music each musician plays a certain instrument or voice (e.g., the bass line, the main melody, the drums), in the piece *Minister of Pitch* each musician is responsible for one aspect of all instruments, of all musicians: one for changes of timbre, one for timing, and another, the "Minister of Pitch," for pitch.

Jacques Attali assumes that music is generally the art form that can most quickly test new social and political models long before they are elaborately implemented in society. The Hub practice precisely this testing of systems, which are always to be heard as expressions of the aesthetic as well as the social and political, when they give up their artistic autonomy and share control of the music with other people and with electronic systems—quite contrary to the great tendency of European art music in recent centuries in which the power of composer and conductor expanded more and more.

What happens if we change the rules? What happens when we couple the net ever more closely to our bodies, our perception, and our society? The Hub is asking these questions by means of a sonifying test bed of possible and possibly impossible socio-technical structures.

17 Jacques Attali, *Noise. The Political Economy of Music* (Minneapolis: University of Minnesota Press, 1985), 5.

ONCE UPON A TIME IN CALIFORNIA

Carter Scholz

I first heard the musicians of the Hub in 1979, shortly after I moved to Berkeley from New York City. They weren't called the Hub at the time; they were the League of Automatic Music Composers. The "hub," the eponymous piece of hardware that would electronically link the musicians, was still in the future. But the linkage was already there: The aesthetic and some of the personnel were the same. The venue was the Finnish Hall, built in the 1930s as the home of a Finnish-American mutual benefit society. Like the "free jazz" played in Soho lofts I had frequented a few years before, this radical music was rooted in mutual benefit and a vision of social change.

As I recall, on that day the personnel were John Bischoff, Jim Horton, Rich Gold, and David Behrman. I knew Behrman's music from a recent recording on the Lovely Music label (*On the Other Ocean*, 1978). The other musicians were new to me. Set up on folding tables was an imposing clutter of arcane gear, including several KIM-1s. The KIM-1 came on the market in 1976 as one of the first hobbyist microcomputers. It had 1K of RAM, a clock speed of 1 MHz, a 16-digit keypad for programming, and 15 I/O lines for connecting external devices. The rest of the gear on the tables looked homebrew.

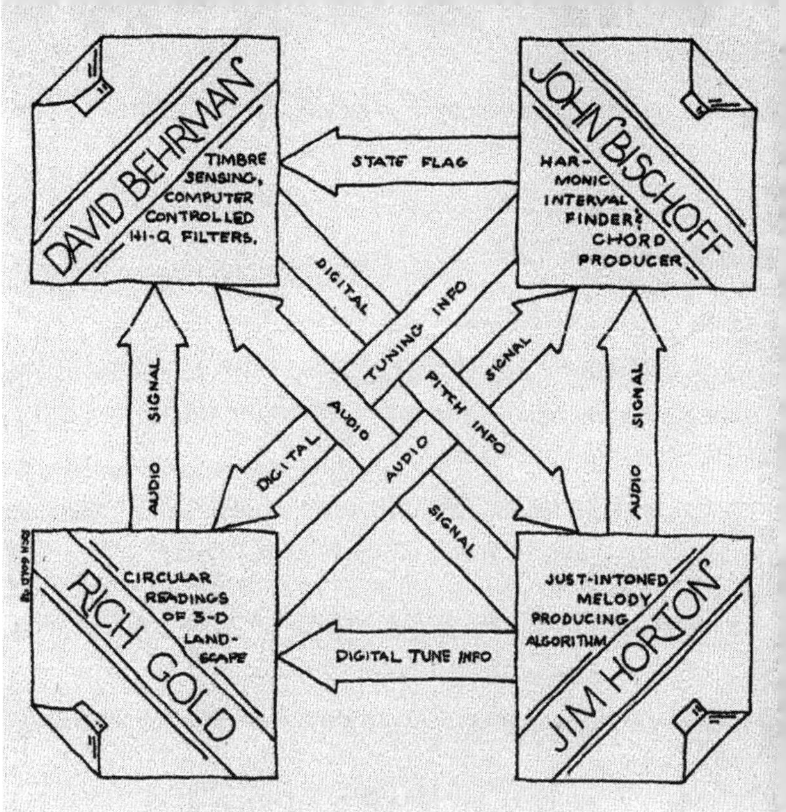

Rich Gold, a founding member of the League of Automatic Music Composers, made this diagram of one of the League's pieces, with enough salient detail that it could serve as a score.

The sound was nothing like Behrman's slow-moving, mellifluous record with its long tones and pure harmonies. It was loud, raucous, abrasive, and compelling. It was music that was, to use the title of Val Wilmer's book on free jazz, "as serious as your life."

In New York I'd heard the young George Lewis play a solo trombone concert at Sam Rivers's Studio Rivbea. At the time, I didn't know that Lewis was interested in electronics and computers, but you could hear it in his purely acoustic playing. I felt a direct continuity between these very different musics.

In Berkeley the energy didn't come directly from the musicians. With electronics, the sound isn't created by your body—though this sound, unlike much electronic music, had a raw physicality. After the sound was set in motion, the players sat back, listened, and occasionally made some small adjustment that would shift the sound minutely or tidally. But the music seemed also to adjust itself, as if by some internal process.

PROCESS

A drawing by Rich Gold shows the process for one League concert. Horton is generating just-intoned melodies by algorithm. Gold is running his "terrain reader," a kind of Martian rover that travels a synthetic landscape, reading elevation to create digital waveforms. Gold is also receiving pitch information from Horton's melodies. Behrman controls high-Q resonant filters with which he modifies Horton's and Gold's sounds, and sends pitch information back to Horton's computer. Bischoff looks for certain pitch intervals in Horton's and Gold's sound, plays chords in response, sends tuning information back to Gold, and information about the intervals to Berhman. These four ongoing processes are autonomous and self-sustaining, yet they influence and affect one another.

The term "process music" probably derives from Steve Reich's 1968 essay, "Music as a Gradual Process," in which he wrote: "once the process is set up and loaded it runs by itself." Reich's emphasis is on readily perceptible process ("I want to be able to hear the process happening"), as opposed to John Cage's processes, which he said were "compositional ones that could not be heard when the piece was performed."[1]

That distinction is perhaps less significant than Reich makes it out to be. Reich's own compositional path—from phase-shifted speech (*It's Gonna Rain,* 1965) to melodic patterns (*Four Organs,* 1970) to tonality (*Music for Mallet Instruments, Voices, and Organ,* 1973)—took him progressively further from readily perceptible process.

Reich also wrote, "one can't improvise in a musical process—the concepts are mutually exclusive."[2] This is a questionable assertion from such a careful writer, even for gradual, perceptible processes, unless one's definition of "improvisation" is extremely narrow. (Cage also had a dim view, and a narrow one, of improvisation.) Indeed, Reich had first-hand experience as one of the performers in the premiere of Terry Riley's *In C,* a piece which allows considerable performer freedom within its process. We'll come back to improvisation.

In any case "process" was very much in the air, as part of the 1960s West Coast rejection of European serialism sig-

1 Steve Reich, "Music as a Gradual Process," in *Writings about Music* (Halifax: The Press of the Nova Scotia College of Art and Design and New York: New York University Press, 1974), 9–11.

2 Ibid.

The League of Automatic Music Composers, (L–R) Jim Horton, Tim Perkis, John Bischoff.

naled by Reich, Riley, La Monte Young, Ramon Sender, and Pauline Oliveros, all young Bay Area composers. Strongly influenced by their examples, the even younger composers of the League had a more expansive, less minimal take on "process," and it included computers.

TECHNO-UTOPIA

In California in the 1970s, microcomputers had a utopian aura. Until then, huge and costly IBM mainframes defined computing; they were the domain of financial and academic institutions and the military-industrial complex. The advent of large-scale integrated circuits made it possible to put a computer, or at least a central processing unit (CPU), on a

3 Ivan Illich, *Tools for Conviviality* (London: Marion Boyers, 2009), 18.

4 D. Schuler, "Community Networks: Building a New Participatory Medium," *Communications of the ACM* 37 no.1 (1994), 38.

5 John Bischoff and Tim Perkis, liner notes to *The League of Automatic Music Composers 1978–1983*, New World Records 80671-2, 2007.

chip, affordably. Personal computers were still in the future, but change was on the horizon. "One user/one CPU" became a '70s battle cry for empowering the individual with this new technology.

This utopia was about more than individual liberation. Microcomputers were to be tools for conviviality, as Ivan Illich put it. "I consider conviviality to be individual freedom realized in personal interdependence and, as such, an intrinsic ethical value."[3]

One such early techno-utopian project was centered in Berkeley. Community Memory was a public-access electronic bulletin board started in 1973 by Lee Felsenstein and others. A teletype connected by modem to a remote time-shared computer enabled the unmediated storage and retrieval of messages, made available to anyone who wished to use it. The makers wrote that "strong, free, non-hierarchical channels of communication—whether by computer and modem, pen and ink, telephone, or face-to-face—are the front line of reclaiming and revitalizing our communities."[4]

These ideals of conviviality, sociability, and interactivity became central to the Hub's aesthetic, and this aesthetic is inseparable from an ethical stance. The "network" of network music is inseparably technological and social. Music is a social activity.

ACTOR-NETWORK THEORY

While there is no chance that any members of the League had read Bruno Latour in 1978 (his first book, *Laboratory Life*, was published in 1979), the band's practice foreshadowed his "actor-network theory" developed in the 1980s. It is less a theory than an ontology, since it focuses on description rather than explanation. It holds that the inanimate or non-human can be co-equal actors in a network.

> *The microcomputers were always just components with particularly interesting behavior to incorporate into our networks, which included other electronic circuitry, as well as human beings.*[5]

Or, as Jim Horton described the League of Automatic Music Composers, "a club with human and non-human members."

If Latour was not yet a touchstone for the group, authors with similarly broad conceptions of systems were: Norbert Wiener (*Cybernetics,* 1948), Ilya Prigogine (*Self-Organization in Non-Equilibrium Systems,* 1977), John Henry Holland (*Adaptation in Natural and Artificial Systems,* 1975), Warren McCullough (*Embodiments of Mind,* 1965), Buckminster Fuller (*Synergetics,* 1975), Gregory Bateson (*Steps to an Ecology of Mind,* 1972), René Thom (catastrophe theory), etc. "Complex phenomena can be understood by analyzing the dynamic interactions of relatively simple components connected in networks."[6]

6 Ibid.

7 Ibid.

CALIFORNIA

Perkis and Bischoff describe the '70s Bay Area milieu as "a rich blend of communal ideologies, radical culture, technical innovation, intellectual ferment, and a hands-on attitude that has been a hallmark of California life since the pioneer days."[7]

The hands-on California composers and inventors would include Henry Cowell, Harry Partch, John Cage, and Lou Harrison, all of whom left marks both technical and social in the Hub's music.

Cowell pioneered inside-the-piano techniques, which Cage adapted into his prepared piano, and which Chris Brown adapted into his Gazamba, a kind of prepared electric piano. Cowell's book *New Musical Resources* presented a panoply of unconventional structural ideas for composition. Cage's early compositional structures show Cowell's influence. The idea of using harmonic pitch relationships as a rhythmic device was incarnated by Cowell in an instrument he called the Rhythmicon, built for him by Léon Theremin.

Cowell's interest in harmonic relationships so influenced Lou Harrison that by the 1950s Harrison came to compose exclusively in various forms of just intonation. Harrison went on, with his partner Bill Colvig, to build American versions of Javanese gamelan, tuned in just intonation, which inspired an "American gamelan" movement in the 1970s and 1980s, in which Tim Perkis participated; the Colvig/Harrison instruments are still used at Mills College.

Magnificat

8 Ibid.

9 Jan Swafford, *Charles Ives: A Life with Music* (New York: W. W. Norton, 1996), 371.

Harry Partch acknowledged no direct influence from Cowell, but he certainly read Cowell's *New Music Quarterly*, and all his mature music was composed for instruments he designed and built himself, tuned in a unique just intonation.

All these composers were self-reliant. They worked outside of institutions. They were all composer/performers, who put together their own performing ensembles. They believed that the function of the composer is to pursue new musical ideas, not audience.

From the beginning, the League and the Hub followed that credo. Bischoff and Perkis:

> *Since the audience was sparse, and opportunities for an actual career futile, why not spend one's efforts following the potential of fantastic ideas, rather than worrying about the practical applications of those ideas within traditional musical domains?*[8]

This declaration of freedom from commerce had deep roots. Charles Ives once demanded that Cowell remove the copyright notice from a piece of Ives's he'd published in *New Music Quarterly*. "EVERYBODY who wants a copy is to have it!"[9]

MILLS & CCM

Mills College in Oakland, California, has long been a hub of avant-garde music-making. A celebratory record, *Music From Mills* (1986), gives a sampling of faculty and students over the decades: Lou Harrison, Terry Riley, Luciano Berio, Dave Brubeck, David Rosenboom, Robert Ashley, Anthony Braxton, David Behrman, Elinor Armer, Steve Reich, Maggi Payne, Darius Milhaud, Pauline Oliveros, Anthony Gnazzo, Katrina Krimsky, Larry Polansky, Pandit Pran Nath, Janice Giteck, Robert Sheff, Ramon Sender, and Morton Subotnick.

The future members of the Hub—John Bischoff, Chris Brown, Scot Gresham-Lancaster, Tim Perkis, Phil Stone, Mark Trayle—were all connected with Mills's Center for Contemporary Music (CCM) in the 1970s and early 1980s. The CCM started life as the San Francisco Tape Music Center, founded by Ramon Sender and Morton Subotnick in 1961—a collective of young, independent composers and artists that included Pauline Oliveros, Steve Reich, Terry

Riley, Tony Martin, the Anna Halprin Dance Company, and many others. It moved to Mills in 1966, where Oliveros was its first director. In keeping with the ideals of conviviality, she insisted that it remain a public-access facility, open to the community as well as students.

The CCM's culture and ethos, inherited from the Tape Music Center, is strongly independent and extra- (almost anti-) institutional, despite its existence within Mills. The young composers were thus exposed to Lou Harrison's gamelan, Robert Ashley's television operas, the tonal minimalism of Terry Riley, the Kirana vocal music of Pandit Pran Nath, the theories and compositions of James Tenney, and much more. Bischoff and Brown went on to join the Mills faculty and to influence generations of their own students.

To this rich stew of influences, add the African American experimental music tradition, from Charlie Parker to Thelonious Monk to Sun Ra to Ornette Coleman and Cecil Taylor and the AACM (Association for the Advancement of Creative Musicians), several members of which later taught at Mills: Anthony Braxton, Roscoe Mitchell, George Lewis.

Most importantly, the CCM brought these students into direct contact with the pioneers of live electronic music. Tudor had been using live electronics in performance with the Merce Cunningham Dance Company since 1961 (his realization of Cage's *Variations II*); Gordon Mumma, Robert Ashley, David Behrman, and Alvin Lucier had formed the ONCE Group, later the Sonic Arts Union, in Ann Arbor in 1961. The idea of using the electronic system itself as a musical actor, as opposed to merely a tool, started with these composer/performers.

TWO STREAMS

If an avant-garde needs a metropole to assert its difference, the metropole in this case would be the academic activity in computer music historically represented by large institutions like Bell Laboratories, the University of Illinois, Columbia/Princeton, and Stanford. Early electronic instruments—the telharmonium (1897), the theremin (1922), the ondes martenot (1928), the trautonium (1930)—emulated the roles of traditional instruments, and later institutional developments followed this example. Research was generally directed towards more realistic synthesis of existing instruments, improved artificial reverberation, compositional algorithms to imitate classical models, and the like.

This approach is teleological. Teleology has a predetermined goal in mind, and directs all effort toward that goal; in this effort, technology is a tool for exerting more power and control towards your end.

The opposite approach is ontological. Technology is not an adjunct or an assistant, but the material itself. This approach seeks new sounds and techniques unique to the new technology. Equipment is repurposed, surprise and unpredictability are welcomed, imperfection and the defects that engineers call "artifacts" are of interest. (The name of the record label created by and for the Hub is Artifact Recordings.) There is an enthusiasm for sounds newly in the world, that sound like nothing before them. The sounds, following Cage, are permitted to be themselves. The search, following Cowell, is for new musical resources.

In the 1980s the teleological approach was exemplified by Stanford's CCRMA (Center for Computer Research in Music and Acoustics). Composers wrote code to generate flawless, timbrally sophisticated pieces. High-end computers would run for hours or days to produce ten minutes of music to be played in a darkened hall with speakers on an empty stage. Thirty miles north at the CCM, chronically short of funding, students would wrestle homebrew electronics and tangles of cable into setups prone to delays, false starts, and outright failures—but it was live. Music is a social activity. Tudor wrote in 1976:

> *Electronic components and circuitry, observed as individuals and unique rather than as servo-mechanisms, more and more reveal their personalities, directly related to the particular musician involved with them. The deeper this process of observation, the more the components seem to require and suggest their own musical ideas, arriving at that point of discovery, always incredible, where music is revealed from 'inside' rather than from 'outside.'*[10]

This is the heart of the ontological approach. Bischoff describes this as "focus on the material nature of the medium."[11]

10 You Nakai, *Reminded by the Instruments: The Music of David Tudor* (London: Oxford University Press, 2021), 303.

11 John Bischoff, "Software as Sculpture: Creating Music from the Ground Up," *Leonardo Music Journal* 1 no.1 (1991): 37–40.

Tudor made extensive use of feedback—putting the outputs of circuits back into their inputs. In its purest form, this produces the uncontrolled yowl of a microphone too close to a speaker. In another form, you have Jimi Hendrix playing the feedback of his guitar pickup by moving it near to his amplifier.

Gordon Mumma's compositions for instruments and his homebuilt "cybersonic" systems also relied on feedback. In *Hornpipe* (1967), a Waldhorn, electronic processors, and the acoustics of the performing space make up one single complex instrument to be explored by the performer. Mumma always emphasized that he was not interested in "control" of his systems, rather "influence."

With feedback, there is a thin, unstable line between yowl and sounds of interest. To explore this line requires virtuosity of an entirely new order. The Hub never made use of direct acoustic feedback, but they create complex data feedback loops within their computer network. They took Tudor's and Mumma's example to a new level.

IMPROVISATION

Complex systems are unpredictable. When you're playing unpredictable music, you'd better know how to improvise.

Val Wilmer writes: "to the Afro-American the freedom inherent in improvisation is his or her birthright." That is not so for white musicians. Although European art music is rich with brilliant composer/improvisers—Bach, Mozart, Beethoven, Schubert, Mendelssohn, Chopin, Liszt, to name only a few—by the twentieth century improvisation had disappeared from European art music (though preserved a while longer in organ pedagogy).

How did this happen? The move from aristocratic courts into public concert halls played a part, as did the availability of printed scores. Notation became more detailed and specific. Figured basses and cadenzas were written out. A rising petit-bourgeoisie associated art music with cultural prestige and created a new market for performers and impresarios, and a new situation for composers. The roles of composer and performer were separated as never before, turning them both into technicians. The nineteenth-century spread of the conservatory, which previously existed only in Italy, hastened the change.

The mid-twentieth-century art-musicians who were misfortunate enough to have inherited this tradition had to reinvent or rediscover improvisation, if they didn't eschew it entirely. This reinvention seemed determined to avoid any contact with the closest living tradition it could have used—African American music, especially jazz. If the prestige-value of this art-music had supplanted its art-value, it is easier to understand why so many white musical technicians shunned anything associated with a socially inferior form of music. After all, European art-music was the child of privileged courts, while jazz was the child of brothels.

But artists are practical. They'll pick up and share anything of use to their art, no matter its origin or associations. The bebop practice of improvising over chord changes was of no direct use to classically-trained musicians, even avant-gardists, but still they looked for some way to escape the tyranny of notation.

It's interesting that "free" improvisation—without any predetermined structure—arises in America at about the same time in the Eurological and Afrological music worlds (to use George Lewis's terms). Pauline Oliveros started freely improvising in 1958 in San Francisco in a group including Loren Rush and Terry Riley. Ornette Coleman's first recorded performances of "free jazz" were made in 1959 and 1960. (Lennie Tristano made earlier forays, but didn't pursue them. In some ways Cecil Taylor also preceded Ornette, but Cecil is a world unto himself.) In Europe, AMM, comprised of Eddie Prévost, Keith Rowe, Lou Gare, Lawrence Sheaff, and Cornelius Cardew, started in 1965. Musica Elettronica Viva, started in Rome in 1966 by American composers Alvin Curran, Frederic Rzewski, Richard Teitelbaum, and others also embraced free improvisation, remaining active, intermittently, for fifty years.

The Hub, coming along later, simply accepted that improvisation, in all its forms, was an indispensable part of almost every musical tradition in the world. It would have been impossible to honor their diverse influences without it. It also would have been impossible to deal with the complex and unpredictable nature of their material without it. Most of all, it was in the air. As Chris Brown wrote of the '80s Bay Area experimental music scene: "Free improvisation was the lingua franca where jazz musicians, junk percussionists, instrument inventors, and computer hackers could come together to play."[12]

And they pointedly called themselves a "band" to indicate how much they owed to the tradition of live improvised music that was kept alive in America by African American musicians.

As it happens, George Lewis saw the League of Automatic Music Composers perform. He said:

> *It sounded a lot like a band of improvising musicians. You could hear the communications between the machines as they would start, stop, and change musical direction. Each program had its own way of playing.*[13]

But there are many kinds of improvisation; it is always specific to its situation. Bischoff writes:

> *Whatever is not defined by the piece spec [specification] is open to improvisation or choice in the moment. In some pieces this difference is clearly delineated, like in* Simple Degradation *where the amplitude for all sounds is determined while all other attributes are open, and in* Tesla Sync *where the potential 'when' of each sound is determined while the exact 'when' and all other features are open. In other pieces the distribution of fixed and open is more evenly integrated across many attributes and a listener in those pieces might feel a global coherence but not be able to locate exactly where correlations are taking place. In either case, improvisation plays a critical role—but the nature of performance 'action' in the Hub is usually shaped so severely by the spec that the feel of an individual voice is often more muted than in other improvised musics. Individual actions are more akin to nudging than declaiming.*[14]

Bischoff is asserting that the structure of a Hub piece is dispositive; even though the musicians have great freedom within the process, it's the process, not the musicians' whims, that defines the piece. This is closer to Cage's "indeterminacy" than to jazz improvisation, but the strong influence of jazz is there all the same.

12 Chris Brown, "Once Upon a Time in California," Radio Web MACBA, https://rwm.macba.cat/en/research/interruptions2-once-upon-time-ca.

13 Curtis Roads, "Improvisation with George Lewis," in *Composers and the Computer* (Los Altos: William Kaufmann, 1985), 75–88.

14 John Bischoff, email, April 7, 2021.

NOISE

Jacques Attali, in his 1977 book *Bruits* (1985 English translation, *Noise*), defines four historical stages of music's political economy: 1) Ritual, which he calls "sacrifice." 2) Adjunct to power, which he calls "representation," in the sense that it represents the interests of the religious or political actors who fund it, enabled by the technology of printing. 3) Commodity, which he calls "repetition," enabled by the technology of recording. 4) And a nascent stage that he calls "composition."

He locates that fourth stage in the free jazz movement of the 1960s, "a profound attempt to win creative autonomy," as exemplified by Chicago's AACM and New York's JCOA (Jazz Composers' Orchestra of America). He concludes that free jazz "failed in its attempt," because "it ran into implacable monetary censorship."[15]

The monetary censorship is true enough, but his dismissal of this new music as a "failure" is refuted by the continuing viability of generations of AACM musicians and bands, including the seminal Art Ensemble of Chicago, which has been active and commercially successful for over half a century, and the deep and abiding influence of this music upon generations of disparate musicians who may or may not identify as "jazz"—including the Hub.

Attali misconstrues free jazz as "a meeting of black popular music and the more abstract theoretical explorations of European music."[16] By the 1960s, "black popular music" was soul and R&B, not jazz, and many free jazz players had little interest in modern European music—while those who did, like Anthony Braxton, found that interest largely rejected by jazz audiences. But Attali is correct that "the refusal to go along with the crisis of proliferation created locally the conditions for a different model of musical production, a new music."[17]

As we've seen, this crisis extended beyond jazz, into the Eurological modernist avant-garde, of which Attali is no fan. He disparages Cage, Boulez, Xenakis, Stockhausen, and Glass variously for scientism, elitism, abstraction, and meaninglessness: "a music without a market imposes itself on an international elite."[18] But he goes on to predict an outcome that sounds premonitory of the Hub's approach:

15 Jacques Attali, *Noise: The Political Economy of Music* (Minneapolis: U. Minnesota Press, 1985), 138.

16 Ibid., 140.

17 Ibid.

18 Ibid., 116.

Music escapes from musicians [...] their role is only to guide the unpredictable unfolding of sound production, not to combine foreseeable sounds originating from stable instruments. [The composer] becomes the organizer of a fluid work: quasi-alive, quasi-organic [...] a work no one except the composer can perform, often for lack of instruments specially invented for it.[19]

19 Ibid., 115.

20 Ibid., 118.

One aspect of Attali's "repetition" stage is that its technology promises perfection; even live performances can be edited after the fact to produce a flawless recording. "This vision gradually leads people to forget that music was once background noise, and a form of life, hesitation, and stammering."[20]

Even punk-rock, which came of age in the late 1970s and early 1980s, finds an echo in the Hub, with its abrasive sound, its explicit rejection of the commercial, its anarchic politics, its rebellion against convention and authority. It is an interesting coincidence that Iggy Pop (James Osterberg) as a teenager attended the ONCE Festival in his home town of Ann Arbor, later acknowledged the influence of Ashley, Cage, and Partch, and once played in a blues band with "Blue" Gene Tyranny (Robert Sheff). But the Hub is more radical than punk in its near-total rejection of idiom. Punk retains the guitars and drums and beat of rock-and-roll; the Hub is suspicious of any idiom, as idiom is the first cousin to cliché.

One of the purposes of the Hub's music-making was to reinject the noise of life into this increasingly sterile world—ironically by using the same technologies perfected to banish that noise. Tim Perkis's 2006 film about Bay Area experimental musicians is titled *Noisy People.* Music is a social activity.

MAKE IT NEW

A serious artist engages with tradition, the history of their art, choosing what to embrace and what to renovate. European art music is formal, so the usual avant-garde response has been not to reject form, but to find new ways of structuring music.

Arnold Schönberg replaced triadic harmony and melodic development with the tone row. Anton Webern and his followers extended the row to the serialization of dynamics and timbre and duration. Ives used juxtaposition as a structural principle. Cowell tried almost everything. Cage at first used numerical forms, then moved to chance processes. Xenakis used stochastics. Reich made gradual process into structure.

The question they all asked is: What parts of music are salient? What characteristics can be used structurally?

Cage wrote that any musical sound has four "parameters": pitch, amplitude, duration, and timbre. Xenakis found "density" to be a useful addition. One could add parameters at will, as taste or need dictates—"harmonic complexity," "rhythmic regularity," and so on.

Borrowed from math, the "parameter" has proved a powerful way of thinking about musical elements, and structuring them. It quantifies qualities, embodying a vaguely Hegelian faith that changes in quantity enable changes in quality.

A parameter usually ranges along a one-dimensional continuum. Pitch ranges from low to high, amplitude from soft to loud, duration from short to long, timbre from simple to complex.

The serialists separated these physical parameters of music from the expressive vagueness of romanticism in much the same way that information theorist Claude Shannon separated "information" from "meaning." In Shannon's formulation, a message with complete predictability (e.g., ABCDEFG) carries no "information." A message of total disorder (e.g., 3hJ+w@1), because of its complete unpredictability, carries a maximum of information. The amount of "information" in a message (i.e., its propensity to inform) is equivalent to its level of unpredictability.[21]

This is counterintuitive because we usually associate "information" with "meaning." But Shannon said, "The 'meaning' of a message is generally irrelevant."[22]

A similar cold-blooded logic seems at work in reducing music to its physical parameters. But this logic was forced upon composers who found that conventional "meanings" in music had been overused to depletion. The remedy was to re-examine the elements. The "meaning" would have to come from somewhere else.

21 Phil Stone, "Non-Mathematical Musings on Information Theory and Networked Musical Practice", *Organised Sound,* in press.

22 Claude Shannon, "Communication Theory—Exposition of Fundamentals" in *Claude Shannon, Collected Papers* (New York: IEEE Press, 1993), 173.

The Hub makes extensive use of parameters. In Tim Perkis's *The Minister of Pitch,* one player controls the pitches played by all, another controls the timing, and a third controls the overall timbre—an overturning of conventional musical practice. Sometimes the parameters aren't even defined. In John Bischoff's *Perry Mason in East Germany:*

> *Each player runs a program of their own design which constitutes a self-sustaining musical process. Players regularly send three bytes to three other players which indicate something about the current state of their activity. Players also design their program to be influenced by three bytes from three other players.*

Nothing is specified about what those variables represent. It is up to the senders, and to the receivers, to sort that out in ways that result in interesting music.

It happens that Phil Stone, as a young man, had a summer job working for Claude Shannon.

SOUND

> *My God! What has sound got to do with music?*[23]
> —Charles Ives

At last we may be in a position to say something, or to hear something, about the Hub's aesthetic.

In the Eurological/teleological framework, "sound" is an imagined goal to be arrived at through detailed notation, obsessive practice, correction, refinement, the elimination of chance and noise. For the ontologist, sound is what emerges from the playing situation, whether that situation is meticulously constructed or ad hoc. The score of Cage's *Variations II,* realized by Tudor for amplified piano in 1961, specifies and controls nothing about how the piece will sound. The sound of the piece is an epiphenomenon of the compositional process and its particular realization.

The spirit of this tradition of experimentalism is well summed up by Alvin Lucier:

> *Some works even resemble experiments in the scientific sense, in that something is discovered during the course of a performance rather than that a preconceived idea or form is brought into being by the will or skill of the composer.*[24]

23 Charles Ives, *Essays Before a Sonata* (New York: W. W. Norton, 1970), 84.

24 Alvin Lucier, *Eight Lectures on Experimental Music* (Middletown: Wesleyan U. Press, 2018), x.

25 Tim Perkis, "Speaking with the Mindless," talk given at the Defragmentation Conference, Darmstadt Summer School for New Music, July 18, 2018.

The Hub's post-Cagean aesthetics still respect the primacy of the process, but in the context of a "band" the result must also "sound good"—whatever that means to the players.

In the jazz tradition, a player's "sound" is paramount—an identity that sounds like no one else. In the Hub's practice, each player has their own uniquely designed instrument, and is expert in playing it—six strong, individual voices.

The great innovation of the network is shared data. It makes another musical dimension available to players, beyond what they can hear. It opens another channel. The data influences their playing in unpredictable ways, but players retain autonomy in shaping their response to it, which in turn influences the network.

The network of players and computers and score together make up a new instrument. Every composition is a newly constructed instrument which the players need cooperatively to learn to play. To play this instrument well demands a new degree of what, in a more conventional setting, would be called virtuosity—not the pianism of a David Tudor, but the sensibility of a David Tudor.

As with Cage and Xenakis and Braxton and the Art Ensemble of Chicago, rigorous structural principles may define each composition, but the character and identity of the piece still emerges from the sensibility of the composer. No matter the method, Cage sounds like Cage, Xenakis like Xenakis.

Although based on radical premises, the music of the Hub is finally about very traditional musical ideals: performance, interaction, balance, how it sounds; listening, responding, adjusting, influencing. Music is a social activity. And that is perhaps where the "meaning" comes back in.

> *In an authentic conversation, we create something together that neither of us were able to create ourselves. Neither of us know where it's going, or what might emerge. This really is the locus of culture creation, it is the place where what is new in human society happens, the place where meaning emerges.*[25]

> *I am concerned primarily with 'system concepts'—configurations that include sound sources, electronic modification circuitry, control or logic circuitry, playback*

apparatus [...] and even social conditions beyond the confines of technology. I suggest that the most important creative aspect of live-performance electronic music technology is not this or that circuit but rather the total configuration itself.[26]

The real musical work of the Hub [...] goes beyond whatever sound producing means we as artists chose in voicing the compositions we made, to the ways in which those individual voices interacted with each other.[27]

For our work to be truly of interest, musicality itself has to be of paramount importance.[28]

26 Gordon Mumma, *Cybersonic Arts: Adventures in American New Music* (Urbana: U. Illinois Press), 44.

27 John Bischoff and Chris Brown, "Indigenous to the Net," HTML files on disc 2 of *Boundary Layer* CD, Tzadik TZ8050-3, 2008.

28 John Bischoff, "Giga-Hertz-Preis 2018: the Hub Lecture/Talk 25.11.2018," ZKM, https://zkm.de/en/event/2018/11/giga-hertz-award-2018-award-ceremony.

THE SCORES

To call the texts presented here "scores" in the context of conventional Western musical practice is to invite a misconception. So let's dispel that misconception quickly.

To "score" is simply to inscribe, to make a mark. When John Cage published *Notations* in 1969, the idea of the "score" was already in such flux that he could present 269 diverse examples of unconventional current music notation practice, which barely skimmed the surface of the ferment. The musicians of the Hub were steeped in Cagean tradition, and were well-acquainted with the text scores of Fluxus (Dick Higgins, La Monte Young, et al.), Pauline Oliveros, Christian Wolff, and Musica Elettronica Viva.

The Hub's preferred word for these inscriptions is "spec," short for specification. As an engineering term, it's accurate enough, but it sells short the musical consciousness and intentions of a group of highly sophisticated and self-aware artists.

These scores have no executive function. They can be extremely terse or extensive. They don't tell a player what or how to play, they don't coordinate actions, they don't prescribe a specific musical outcome. They are not recipes. They establish a situation for music-making. The situation for music-making that the Hub prefers is one of openness, unpredictability, surprise—a situation from which a musical result emerges that would not have occurred otherwise.

Phil Stone: "The notion of 'surprise' comes up again and again in Hub discussions. Successful pieces are described as having the capacity to delight with unexpected or emergent behaviour. Over-specification or excessive demands for conformity are viewed with suspicion for their potential to damage or destroy this capacity."[29]

As might be expected in a working band, the development of a piece only begins with the spec. Many details were worked out in email and discussion and rehearsal and performance, and not necessarily documented. The written spec of a piece is unlikely to reflect all this give and take. Wherever possible and appropriate, I have included comments to give context and to clarify and explicate aspects of pieces that might be obscure in the spec.

To me, the main interest in the scores is in seeing how the Hub thinks about music, and hearing how that plays out in practice—which are two separate things. Listening to the music with the score in mind may be enlightening, but becoming fixated on either the Hub's sound, or its particular methods, may miss the point.

Nonetheless it should be possible for another group of computer musicians to realize a Hub spec from what is presented here. Gresham-Lancaster's *Stuck Note* has already been used to teach network music concepts. But would such realizations sound anything like the Hub's own version of the same piece? Should they?

Phil Stone: "Most pieces could be adopted to a huge variety of techniques and/or technological solutions."[30] Throughout its history, the Hub was constantly updating its technology. No two players ever shared the same software/hardware combination.

John Bischoff: "There is no Hub 'style' in any traditional sense. The particular sounds and structures in the music are a unique experiment for each piece."[31] So each piece has its character, a character that gradually revealed itself to the Hub as they rehearsed and performed it. But that character might be revealed in a wide variety of sonically different ways.

As Chris Brown says, the scores "are there to be implemented by anyone who would like to use them."[32] Those who would like to do so should be ontologists, with strong

29 Phil Stone, "Non-Mathematical Musings on Information Theory and Networked Musical Practice", *Organised Sound,* in press.

30 Ibid.

31 John Bischoff, "Giga-Hertz-Preis 2018: the Hub Lecture/Talk 25.11.2018," ZKM, https://zkm.de/en/event/2018/11/giga-hertz-award-2018-award-ceremony.

32 Chris Brown, "Giga-Hertz-Preis 2018: the Hub Lecture/Talk 25.11.2018," ZKM, https://zkm.de/en/event/2018/11/giga-hertz-award-2018-award-ceremony.

individual voices, and a commitment to the roots of this music and to their present moment and situation. They should respect the character of each piece, and find their own ways of realizing and revealing it. They should understand that the music is not the sound.

All Hub compositions have a structural spine or strategy that tends to upend conventional musical structure, and this is finally where the piece's character lies. Exemplary is John Bischoff's *Perry Mason in East Germany.* The spec says merely that each player "writes three bytes to the Hub [...] which indicate something about the current states of their activity," and "reads three bytes from the Hub [...] to be influenced by one variable from each of three other players." This quite purposely does not define or suggest in any way what those variables or influences might be. And yet, "the complex web of mutual influence gives the music structure beyond any individual's planning." Remarkably, it does, and the structure is audible.

John Bischoff: "I've often thought of network music as a kind of expanded polyphony where instead of multiple melodic lines that form harmonies, you have multiple musics aligned through diverse parameter correspondences: a sonic fabric knit together of independent musical personalities. It is as if each Hub part declaims its own idea of what music is, and it is the interaction between these declarations—the agreements, disagreements, oblique agreements, interruptions, merging, diverging, ignoring, cajoling, attending—and the resulting musical shapes and contours, that are the focus of the form. It is also important to allow for non-structure as you listen to the Hub. Only then can we appreciate the coalescing of momentary structures within the music and the surprising instances of odd correlation."[33]

33 John Bischoff, "Free Association: Snapshots of an Electroacoustic Musical History," keynote address at Music in the Global Village, Budapest, September 8, 2007.

In the end, the structure may be, as James Joyce said of the structure of *Ulysses,* a bridge to march your troops across, and who cares what happens to the bridge afterwards. The life is in the crossing. The life of that moment.

ACKNOWLEDGMENTS

I want to thank all the members of the Hub for making themselves and their archives available to me. I borrowed the title of this essay (or is it stealing?) from Chris Brown's podcast "Interruptions #2" for Radio Web MACBA.[34]

34 Chris Brown, "Once Upon a Time in California," Radio Web MACBA, https://rwm.macba.cat/en/research/interruptions2-once-upon-time-ca.

FURTHER READING/VIEWING

Ashley, Robert. *Music With Roots in the Aether*, New York: Lovely Music, 1976.

Austin, Larry and Douglas Kahn. *Source: Music of the Avant-Garde, 1966–73*, Berkeley: U. California Press, 2011.

Bernstein, David. *The San Francisco Tape Music Center*, Berkeley: U. California Press, 2008.

Cage, John. *Silence*, Middletown: Wesleyan U. Press, 1961.

–. *A Year from Monday*, Middletown: Wesleyan U. Press, 1969.

–, ed. *Notations*, New York: Something Else Press, 1969.

–. *M*, Middletown: Wesleyan U. Press, 1973.

–. *Empty Words,* Middletown: Wesleyan U. Press, 1979.

–. *X*, Middletown: Wesleyan U. Press, 1983.

–. *I–VI*, Cambridge: Harvard U. Press, 1990.

Cowell, Henry. *New Musical Resources*, New York: Alfred A. Knopf, 1930.

Harrison, Lou. *A Music Primer*, New York: Edition Peters, 1971.

Lewis, George. *A Power Stronger Than Itself: the AACM and American Experimental Music*, Chicago: U. Chicago Press, 2008.

Lucier, Alvin and Douglas Simon. *Chambers*, Middletown: Wesleyan U. Press, 1980.

Lucier, Alvin. *Music 109: Notes on Experimental Music*, Middletown: Wesleyan U. Press, 2012.

Mumma, Gordon. *Cybersonic Arts: Adventures in American New Music*, Urbana: U. of Illinois Press, 2015.

Nakai, You. *Reminded By the Instruments: David Tudor's Music*, London: Oxford University Press, 2021.

Partch, Harry. *Genesis of a Music*, New York: Da Capo Press, 1974.

Perkis, Tim. *Noisy People*, Albany, CA: Poikusfilm, 2006.

Reich, Steve. *Writings about Music*, Halifax: The Press of the Nova Scotia College of Art and Design and New York: New York University Press, 1974.

Tenney, James. *Meta+Hodos and META Meta+Hodos*, Oakland: Frog Peak Music, 1987.

Zimmermann, Walter. *Desert Flowers*, Vancouver: A.R.C. Publications, 1976.

ALPHABETICAL INDEX OF COMPOSITIONS

THE SCORES

HARDWARE/SOFTWARE

Each Hub member always chose his own hardware, software, and coding practices. Nothing was shared in common but the hub itself. This variety, imposed by necessity, turned out to be a strength, as it meant each musician had a strong, individual voice in the whole. Over the years, the computers have included Commodore 64s, Amigas, IBM PCs, Macintoshes, Windows laptops, and MacBooks. Music programming languages included many dialects of Forth, FORMULA (by David Anderson), HMSL (by Phil Burk, Larry Polansky, and David Rosenboom), PureData (by Miller Puckette), Keykit (by Tim Thompson), Max/MSP (by Miller Puckette and David Zicarelli), and Supercollider (by James McCartney and others). Sound-making apparatus included various homemade and commercial analog and digital synthesizers, and in later years software instruments built in PureData, Max/MSP, and Supercollider.

DIAGRAMS

Each score is accompanied by a dataflow diagram, showing how data is shared among players. A round point represents a player, with lines and arrowheads indicating the data flow. For example, the diagram for *Simple Degradation* shows that all data originates from one player (Mark Trayle) and flows to the other five. The diagram for *Borrowing and Stealing* indicates that every player is sending data to every other player. The diagrams were designed and rendered by Judy Coates Perez.

The QR codes in this book link to online audio or video examples of many of the scores. Simply scan the corresponding code to experience the Hub in action. The linked materials can be found collected here: zkm.de/thehub.

THE HUB : COMPUTER NETWORK MUSIC

AT TWO LOCATIONS SIMULTANEOUSLY

THE CLOCKTOWER	&	THE EXPERIMENTAL INTERMEDIA FOUNDATION
108 LEONARD ST.		204 CENTRE ST.
NEW YORK, NY 10013		NEW YORK, NY 10013

SATURDAY & SUNDAY, JUNE 6 & 7 , 1987 at 9 PM

FEATURING:

Chris Brown **John Bischoff**
Scot Gresham-Lancaster **Tim Perkis**
Phil Stone **Mark Trayle**

THE HUB is a computer music band whose members are all designers and builders of their own hardware and software instruments. The group coordinates electronically the activity of their individual systems through a central microcomputer, The Hub itself, as well as manually through ears, eyes, and hands. For this concert the Hub was commissioned to build a new instrument which would allow simultaneous co-ordinated performances of the same piece by the group, split in half, in two different spaces well out of earshot from each other. This is the premiere performance by the group using this new system.

- I -

At both spaces simultaneously

VAGUE NOTIONS OF LOST TEXTURES

Concept by Scot Gresham-Lancaster, inspired by and dedicated to Jim Horton. The Hub is used for the six performers to send text messages to each other, which will describe their transit through a simple musical shape, in this case a sustained and even increase of density, loudness, and brightness of sound. The two groups conduct each other by sending messages across the network, coordinating their positions with regard to the shape, and keeping each other in time. The audience is invited to leave their seats and browse the messages appearing on each performer's computer monitor.

IS IT BORROWING OR STEALING? (II)

Designed by Phil Stone. Hub memory is sliced into six areas which contain the melodies currently being played by six computer/performers. Players borrow or steal melodies from each other at will, sometimes transforming them with transpositions, inversions, time deformations, etc., and always writing what they are doing into their area of the Hub. The changing melodic texture is the result of the borrowing or stealing process. The title reflects on the question: is information (and software) free to be borrowed, or owned to be stolen?

SIMPLE DEGRADATION

Designed by Mark Trayle. One performer generates and processes a waveform, simulating the response of a plucked string. This waveform is then broadcast on the computer network, the other performers using it for amplitude modulation (loudness variation). The rate at which the waveform is played back by the performers is determined by the performer who generated the waveform.

INTERMISSION

There will be time for the audience to move to the other space, if they wish, during the intermission. The second half will have different programs in each space.

HUB 0.0

EARLY PIECES

These early pieces do not have surviving specs. The descriptions are taken from the liner notes, written after the fact.

The first hub used the serial ports of the KIM-1 to communicate between computers at 300 bits per second. Each player had 84 bytes of storage. Perkis and Bischoff used this method when they were playing as a duo, and as a trio with Mark Trayle, before Brown, Stone, and Gresham-Lancaster joined.

TUDOR

John Bischoff and Tim Perkis 1986

Released on *Hub Music* and *Boundary Layer*

Tim conjures David Tudor-esque modulated swells using his Buchla Music Easel on full throttle spring-reverb while John combines sustained chords in 31-tone equal temperament, produced on multiple SID chip oscillators (the 6581 Sound Interface Device), which are processed through a performance-tweaked delay pedal.

HUB multichannel mailbox control program.

Devices connected to each channel make requests to write to the nub processor table memory, and to read it. Each makes its request by sending command bytes of which the high four bits form a command field (CF) and the low four a data field (DF). In the nub processor there are three variables kept for each channel: a current WRITE.ADDRESS,(12 bits); the current READ.ADDRESS, (12 bits) and the current WRITE.DATA (8 bits) . These variables for each channel can be set only by commands from that channel. All channel commands are dedicated to setting these variables, or initiating a read or write to the hub table memory.

COMMANDS

CF	name	function
0	Wadlo	set low 4 bits of write.address= DF
1	Wadmid	set middle 4 bits of write.address = DF
2	Wadhi	set high 4 bits of write.address = DF
3	xdlo	set low 4 bits of write.data = DF
4	xdhi	set high 4 bits of write.data = DF; then write contents of write.data to hub table memory at write.address
5	incwad	increment write.address by DF
6	Wrtnext	set high 4bits of write.data = DF; write hub table memory at write.adr with write.data increment write.address by 1 ;
7	--	
8	Radlo	set low 4 bits of read.address = DF
9	Radmid	set middle 4 bits of read.address = DF
A	Radhi	set high 4 bits of read.address = DF
B	Read	transmit contents of hub table memory at read.address to channel.
C	Rdnxt	Do read; then increment read.adr by one.

HUB MUSIC

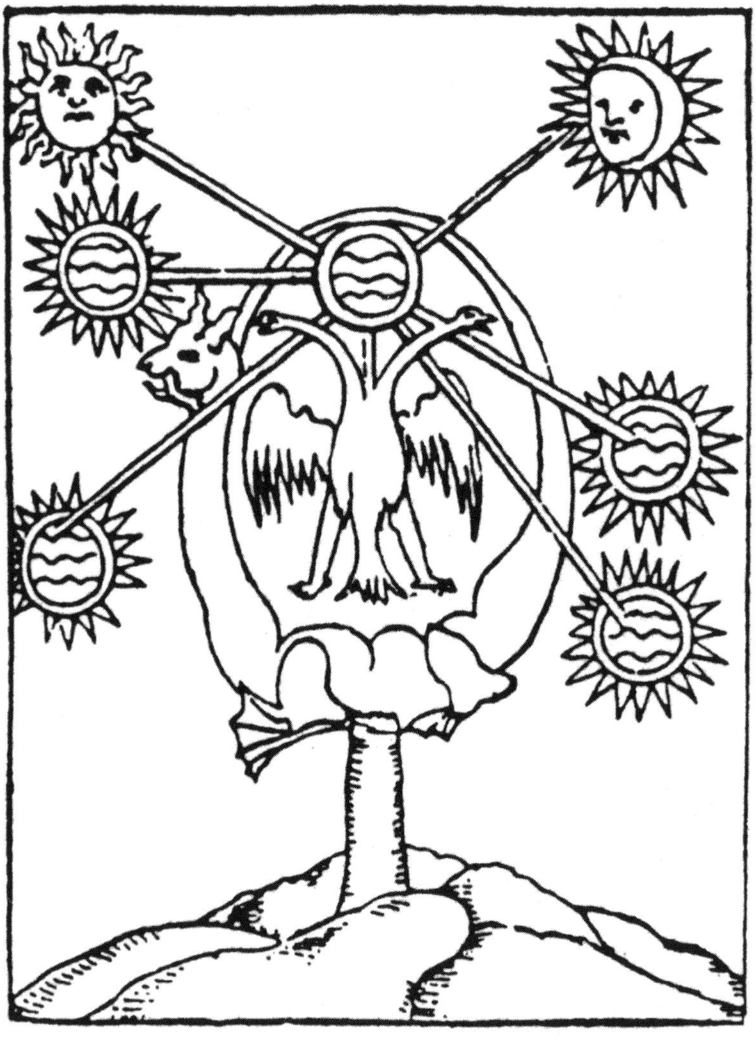

at **The Lab**
1805 Diversadero
San Francisco, Ca.

Nov. 6 & 7 @
8:30 p.m.

WAVE JUNCTION

John Bischoff and Tim Perkis 1986

Released on *Hub Music* and *Boundary Layer*

Two voices of John's "Audio Wave" 8-bit waveform generator play intervals and glissandi in first slow, then rapid, succession eventually evolving into rapidly looping textural states. Tim moves along in tandem, ornamenting sonic motions by sweeping the CPU clock of his hacked Casio.

PEYOTE

John Bischoff, Tim Perkis, and Mark Trayle 1986

Released on *Hub Music* and *Boundary Layer*

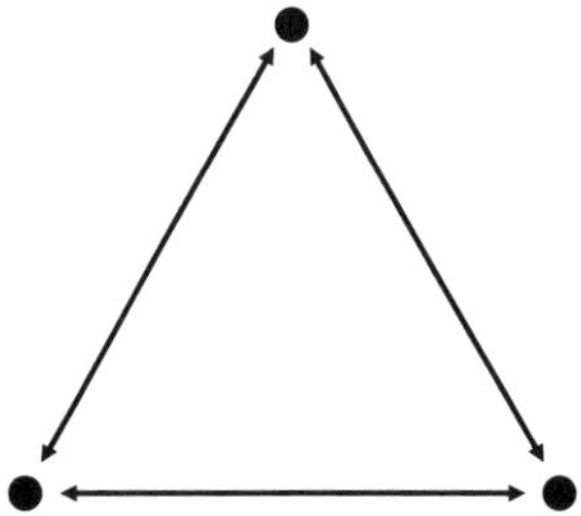

Rhythmic coordination coupled with partial tracking of pitches form the core organizational principles of this piece. Partial linkage between players occurs, a unique brand of musical cohesion investigated by the group at this early stage. The coordination in rhythmic and pitch domains is clearly apparent—but as if the players were aware of each other only at a distance. There is a heterophonic quality to the ensemble texture.

VINI

John Bischoff, Tim Perkis and Mark Trayle

1986

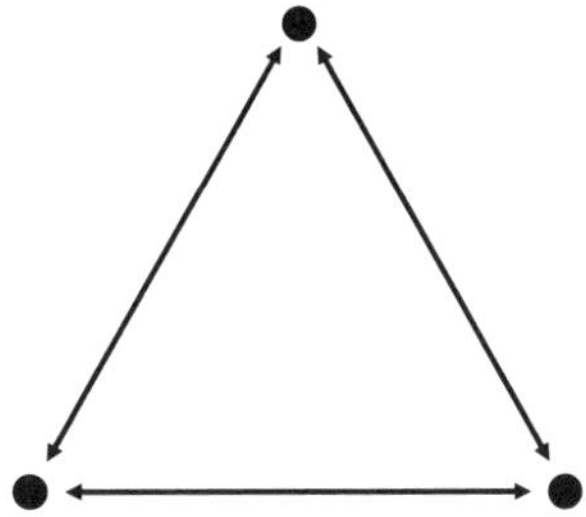

Released on *Hub Music* and *Boundary Layer*

Perkis: "The electronic instrumentation in this early period of the Hub consisted of a distinctive hybrid of analog and digital sound-generating means. Tim made extensive use of Don Buchla's 'Electric Easel' portable analog synthesizer driven by control voltages from his IBM PC. John incorporated his 'Audio Wave' 8-bit waveform generator heavily processed through an analog band-pass filter and tape delay system. One characteristic of these home-brew voices is their mechanistic commonality which allows each to imitate the other in quality and function. The continuous interplay between players that are frequently trading roles—from foreground soloist to textural landscapist and back—is a hallmark of the music."

FALLING EDGE

Chris Brown and Mark Trayle

August 3, 1986

Released on *Boundary Layer*

From Chris and Mark's performance in The Network Muse–Automatic Music Band Festival in San Francisco in 1986, the title refers to the simple trigger that interconnected their homebrew, computer-controlled analog signal processing equipment. Both performers used Serge Modular circuits to modify the sounds of their electroacoustic instruments.

FARABI

Tim Perkis
1986

Released on *The Hub* and *Boundary Layer*

An homage to the Arabic musician and philosopher Abu Nasr Al-Farabi (c. 850–950), this live performance piece was written for a microcomputer system with five analog performer inputs and MIDI. The MIDI sound-generating device is tuned to 31-tone equal temperament, a system of dividing the octave into thirty-one equally spaced steps instead of the usual twelve. The performer's actions regulate tempo, rhythmic regularity, mode, tessitura, and dynamics. In this performance, the MIDI data stream is also sent to Scot's computer/synthesizer system, which is used to rhythmically coordinate his free analog synthesizer playing.

Scot notes: "I was getting absolutely hammered with MIDI data from Tim when we did this and I was playing an Oberheim Xpander 6-voice synthesizer. As the notes came in, the stream was often too low or high, so part of my playing involved active octave switching to stay in an audible range. Also, the Xpander has a rotary encoder that can be used to change all six filters from lowpass to highpass to bandpass very quickly. So I was manually playing the resonance of the intonation by quickly adjusting the filter type and resonance."

HUB × 6

Bischoff, Brown, Gresham-Lancaster, Perkis, Stone, and Trayle 1986

Released on *Hub Music*

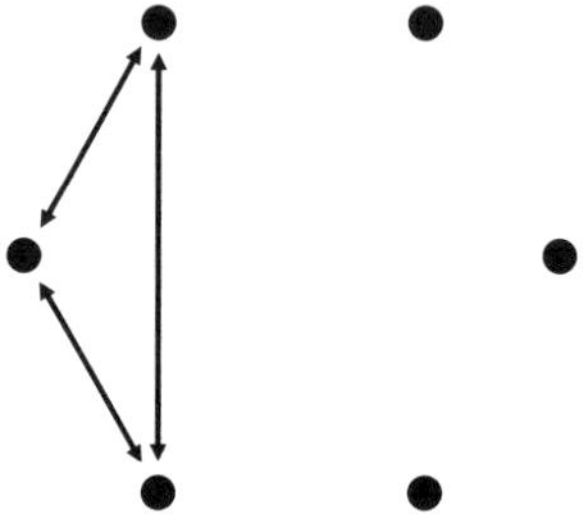

This piece is the first recording of all six members of the Hub playing together. It took place sometime shortly after the Network Muse Festival (see note for *Falling Edge*), and before a hub that could accommodate all six members of the group had been built. It probably involved some networking using the 3-input hub that was built and premiered by Perkis and Bischoff at that festival, with the non-networked members of the group improvising freely.

HUB 1.0

The next version was a pair of homebuilt devices designed by Tim Perkis and programmed by Phil Stone—a memory space where players could deposit and retrieve data relevant to what they were playing. The low communication speed of this device (2,400 bits per second), along with the polling used for communication (data waited, as in a mailbox, for a performer to retrieve it) meant that tight rhythmic coordination was impossible. The shared data referred to other, slower-moving musical parameters, such as pitch and density.

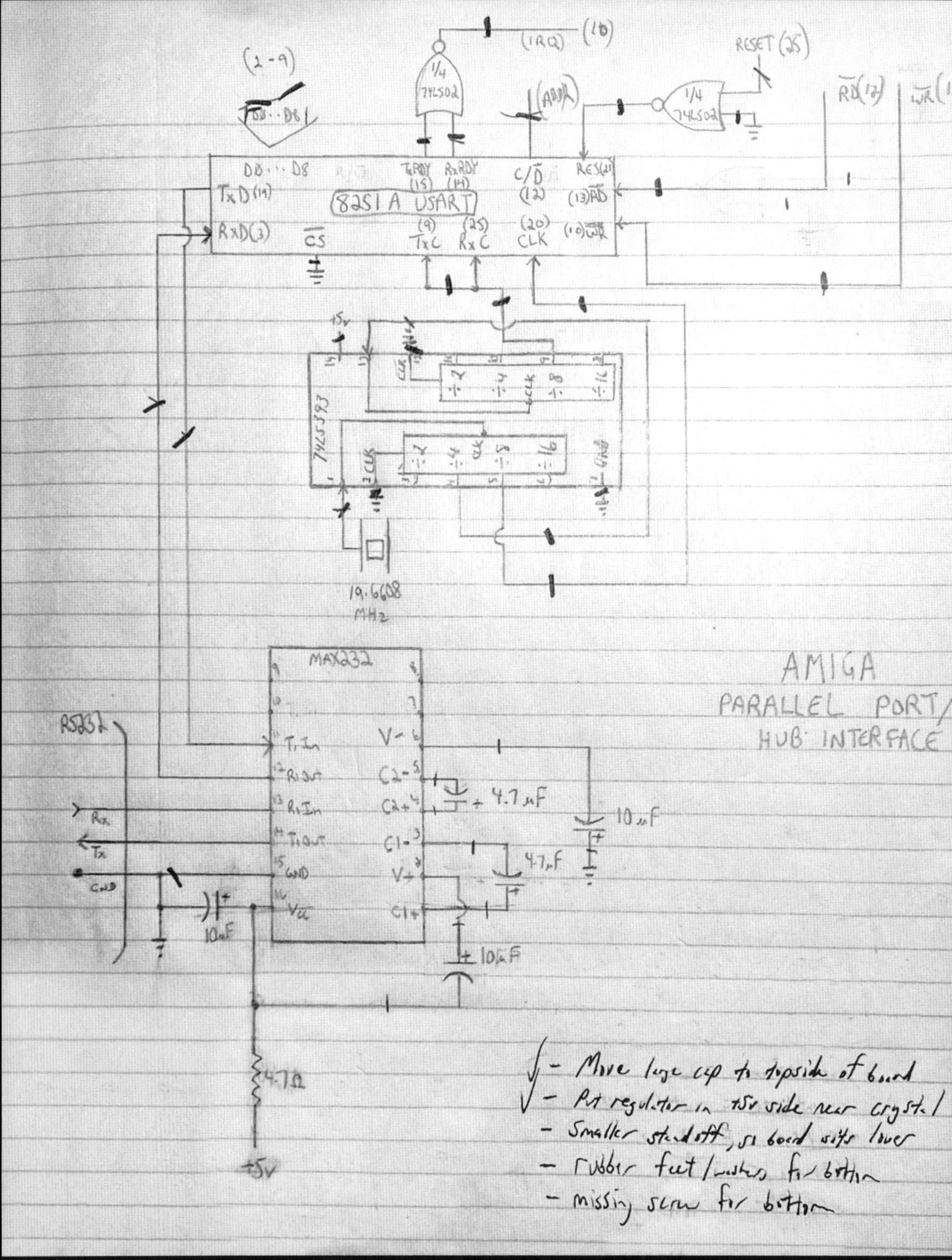

Phil Stone's custom circuit for connecting his Amiga's parallel port to the early hub.

SIMPLE DEGRADATION

Mark Trayle
1987

Premiere: June 6, 1987 at Experimental Intermedia/ The Clocktower, New York Released on *The Hub* and *Boundary Layer*

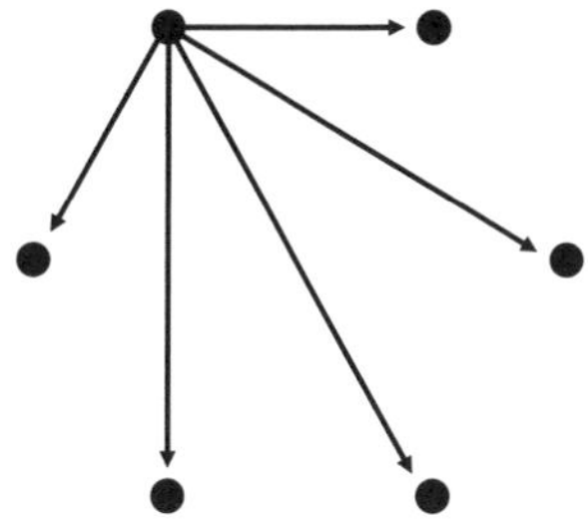

One performer generates and processes a waveform, simulating the response of a plucked string. This waveform is then broadcast on the computer network, the other performers using it for amplitude modulation (loudness variation). The rate at which the waveform is played back by the performers is determined by the performer who generated the waveform. In the Hub version, the performers are free to choose whatever timbres and pitches they wish.

RULES:
The waveform may only be used for amplitude modulation. Pitch may only change after one complete cycle of the waveform.

SPECIFICS:
The waveform is 250 bytes long. It begins at offset 100h for Zero Chat Chat, offset 200h for the MIDI guys, in Hub memory. The sample delay bytes are allocated as follows:

```
01FAh John
01FBh Tim
01FCh Mark
02FDh Scot
02FEh Chris
02FFh Phil
```

The first full Hub piece, written for their first sextet performance at Experimental Intermedia / The Clocktower in New York in 1987. Two trios were physically separated at the two sites, linked by phone lines. Zero Chat Chat was the trio of Bischoff, Perkis, and Trayle running KIM-1s and idiosyncratic analog/digital synthesizers; Gresham-Lancaster, Brown, and Stone were "the MIDI guys" running MIDI synthesizers under computer control.

The data being broadcast on the network is the output of a Karplus-Strong algorithm, a series of numbers that represent the amplitude decay of a plucked string. This amplitude envelope is applied to the freely-chosen sounds of the performers, resulting in a sextet of overlapping decrescendos. Typically a Hub performance of this piece would involve several such "plucks," initiated by Trayle.

BORROWING AND STEALING

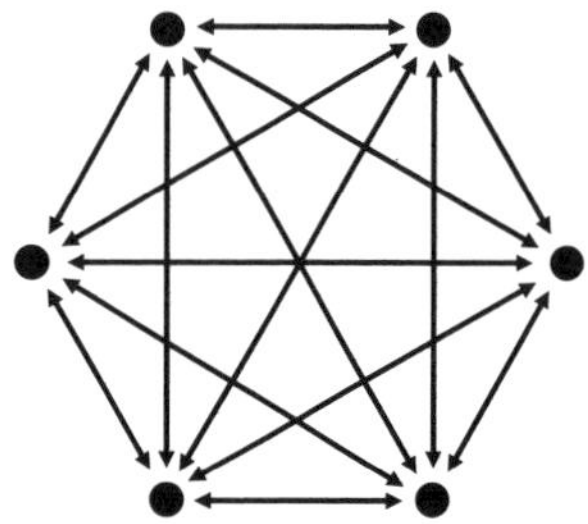

Phil Stone

May, 1987

Premiere: June 6, 1987 at Experimental Intermedia/ The Clocktower, New York

Released on *The Hub* and *Boundary Layer*

SUMMARY

Each performer will be capable of playing a repeating one- to 24-note monophonic melody, using a pre-agreed tuning system. Pitch and duration information, as well as a loop-start synchronization flag, will be sent to the Hub as the melody plays. Each player has their data area (51 bytes apiece: 25 bytes each for pitch and duration values, including a loop-termination "pitch" value, and 1 byte for the loop-start semaphore). Each performer may, at any time, copy information from another performer's melody data and use it for their own melody data.

The piece will start when all performers are ready. The method of ending it will be determined by experiment in rehearsal.

COMMENTS

The only constraint that is placed on how shared data is to be used is that its original melodic or rhythmic information-type be vaguely preserved. For instance, a performer may decide to copy only every other note from another performer's melody, or use only the rhythmic data, or transpose all notes by some interval (whatever!). However, it would not be in the spirit of the piece to use pitch data to control timbre or amplitude modulation. A multitude of complex interactions will be possible just by sharing melody information, without remapping parameters.

Possibilities are much more fun to discuss than constraints. The minimum requirement of a performer's system for participation in this piece is that it be able to play a looping melody of 24 notes or less, and communicate information about that melody to the Hub, in a simple pitch/duration format. How that melody comes into existence is wide open. The performer may choose to pre-compose melodies and play them by manual selection, or generate them live (as long as their system is capable of constructing a loop from live performance).

MISCELLANEOUS NOTES

Timbre: each composer is encouraged to use any method conceivable to actually generate the notes of this piece.

Dynamics: amplitude of individual notes or overall dynamic variation may be manipulated if desired, but this is not required.

DATA FORMAT

Each composer's data area will consist of 51 bytes, with the following significance:

Byte 0: start-of-loop semaphore (set to `$FF` throughout duration of the first note in loop, set to `0` otherwise)

Bytes 1–50: Pitch/duration coordinates for each note in the melody:

Pitch byte: EITHER: a value in the range `0 - 127`, with middle C = `60` (decimal), as in the MIDI specification. Unlike MIDI, a pitch value of `0` indicates a rest, which may be used for articulation as well, if desired. OR: `$FF`, indicating "End of Loop." Duration byte may be ignored in this case.

Duration byte: time, in 1/60 sec., hub timer-derived units, given as the number of these units that the note will sound. The next note in the list is played at the end of this period (use note value = `0` if a rest or articulation is desired).

1 John Bischoff and Chris Brown, "Indigenous to the Net," HTML files on disc 2 of *Boundary Layer* CD, Tzadik TZ8050-3, 2008.

2 Ibid.

A seminal piece for the Hub, also composed for the Experimental Intermedia/Clocktower concert, *Borrowing and Stealing* defines a complex of interactions that actively involves all the players—"all-to-all" data flow, rather than "one-to-all" as in *Simple Degradation.*

Stone typically started the piece by generating a melody which the others would transform. "This could have led to a very formal, tightly controlled music made entirely out of subtle variations [...] but instead there was a more anarchic response as our musical proclivities favored radical transformations [...] Realizing a good performance of the piece became a balancing act between revealing the organization process at its core without making it too dominating of the music [...] Our renditions of the piece probably reflected a creative internal tension in the band between its minimalist and improvisational tendencies."[1]

It's impossible not to read the title as a dry but pointed commentary on "intellectual property," a subject of controversy at the time. Sampling technology was newly available to musicians, with rap artists borrowing (or stealing) riffs and basslines from existing recordings. The composer John Oswald wrote an essay in 1985, "Plunderphonics, or Audio Piracy as a Compositional Prerogative," and released his album *Plunderphonics* in 1988, made entirely of sampled and manipulated recordings, upon which he was sued by Michael Jackson. (A few years later, Jackson himself was sued by the Cleveland Orchestra for unauthorized sampling of one of their recordings.)

Bischoff and Brown: "Its subject and title anticipated what would eventually become the battleground for networked musical technologies—the plunderphonic reality that digital information as symbolic representation of sound is all too easily and perfectly replicated, that such information cannot really be owned, but to be kept alive must be continually transformed."[2]

VAGUE NOTIONS OF LOST TEXTURE

Scot Gresham-Lancaster 1987

Premiere: June 6, 1987 at Experimental Intermedia/ The Clocktower, New York

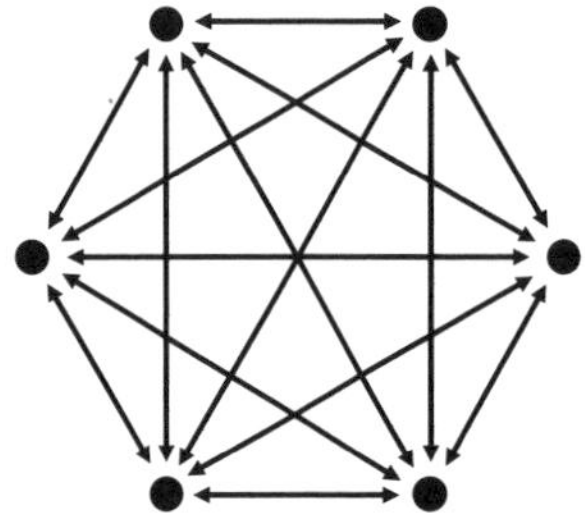

The point of this piece is to increase the density, loudness and brightness of timbre evenly as per the diagram. I would encourage the participants to use sounds with long sustaining timbres so that listeners cannot discern the change of density and sound. I don't care if the sound producing devices are directly controlled by the computer. It is my intention that the Hub be used as a source of text information to convey where each player is in terms of the maximum point we plan to reach at the zenith. The text can be used to coordinate our position in time and density. By looking at the various shared memory locations we can discern where we all are. There will be a main "conductor" clock and density location that is automatically advancing as a point of reference for us to coordinate against.

Inspired by and dedicated to Jim Horton.

Vague Notions implements a simple process that centers on the paramaterization of musical sound—density, brightness, and amplitude are the only parameters specified, with all other decisions left to the performers. The gradual increase of parameter values shown in the sketch is not part of the actual data flow; in performance, the only data exchanged were text messages among all the players, such as "I just brought in the third voice," "We're not really gelled yet," and "We should slow it down." As the composer puts it, "The instruction for this piece is essentially 'make music while carrying on a chat with the other players.'"

Gresham-Lancaster: "Jim Horton was an amazing individual full of rich and colorful ideas regarding the future of music. In a conversation he suggested that musicians should use technology to coordinate and chat with each other to help shape the performance interactively and collectively. I took this idea and fashioned this concept from that conversation."

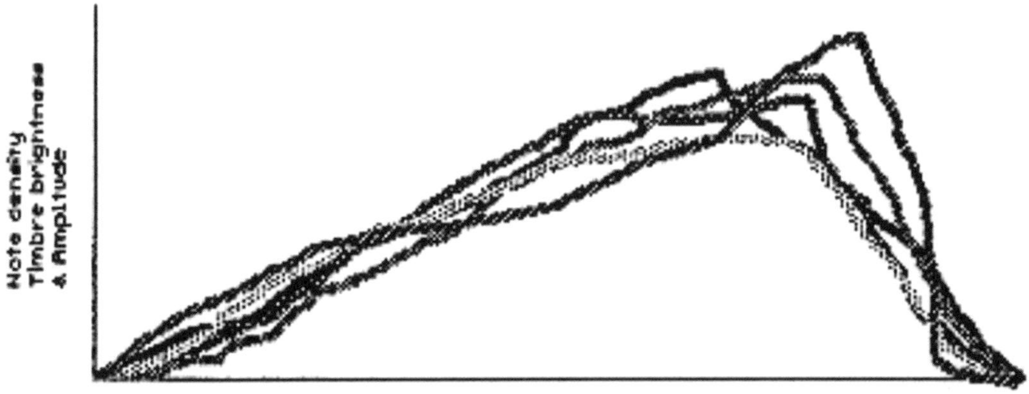

Coordinate system, ***Vague Notions***, 1987.

ROL'EM

Chris Brown
October 19, 1987

Premiere: November 6, 1987 at The Lab, San Francisco

Released on *The Hub* and *Boundary Layer*

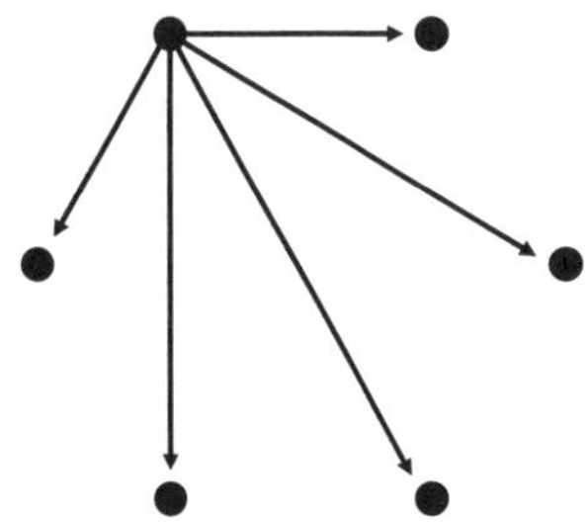

Rol'em defines musical parameters as both performative and acoustic. Thus, for example, a player's "role" might be boss (leader), lead, accompanist, filler, wallflower, or submissive. "Ensemble" is a special parameter, computed in a more elaborate way—e.g., if QUART is chosen first, then one player is sent TACET and the other four are sent QUART; if SOLO is chosen first then all but one player are sent TACET; and if TACET is chosen first everyone gets TACET and the ensemble plays acoustically. All the other parameters are purely random choice, so there can be any number of BOSSes, LEADs, etc.

Explicit theatricality was part of the piece's intention, reflecting influences including John Cage (chance methods), Butch Morris (hand-cueing system), and John Zorn (game structures). It also adds acoustic sound to the group's electronic sound palette, a role that was often taken by Gresham-Lancaster on trombone.

A conducting computer (the conductor) composes musical textures by randomly choosing one from a set of six descriptors of five ensemble parameters:

```
1 Ensemble
  (TACET – SOLO – DUO –
  TRIO – QUART – QUINT);
2 Role
  (BOSS – LEAD – ACCOMP –
  FILL – WLFLR – SUBM);
3 Tune
  (CHROM – 8NOTE – 6NOTE –
  4NOTE – 2NOTE – DRONE);
4 Range
  (HI – MEDHI – MED –
  MEDLO – LO – ALL);
5 Timing
  (GALLOP – TROT – WALK –
  CRAWL – DRIFT – SNOOZ)
```

The conductor sends a set of these choices as text messages simultaneously to each of five computer players who improvise following the parameter descriptors they receive, displaying them on their monitors so they can see them at a glance. The conductor transmits `clock` messages through the Hub to synchronize transitions between changes. It also provides a downbeat clock-time specifying when the newly received descriptors become active. Section durations are between 10 and 60 seconds.

Each player provides a mono signal of unadulterated sound from their "axe" to the conductor. The conductor connects them to a signal processing system where they are adulterated, and otherwise shifted in spectrum and space (mixed).

TWO ADDITIONAL RULES

[1] while playing the role of BOSS or LEAD, the player must stand up while playing, like in Big Bands.

[2] If not playing in the current group (ENSEMBLE = TACET), players are free to play any other acoustic soundmakers they wish (bells, whistles, washboard, sousaphone, whatever).

SPRAY AND ROLL ON

Tim Perkis 1987

Premiere: November 6, 1987 at The Lab, San Francisco

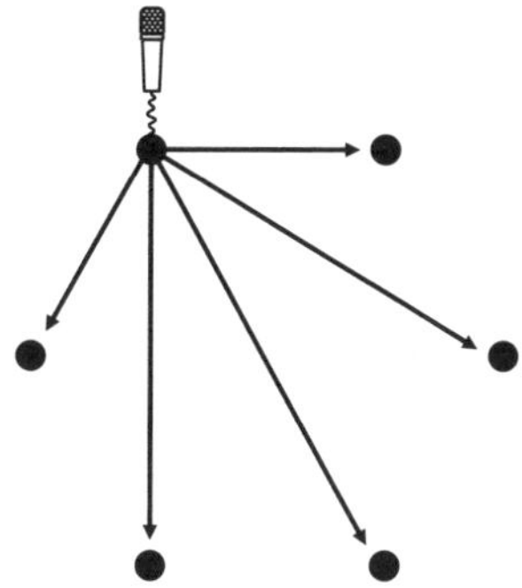

6 to 9 minutes long.

For microcomputer network band (the Hub) and instrumentalist.

One computer system attached to the Hub is designated recorder and CONDUCTOR.

The playing of an INSTRUMENTALIST is recorded through a pitch-to-MIDI converter attached to the CONDUCTOR computer, which then integrates the INSTRUMENTALIST's sequence of pitches into a data area in the Hub which is accessible to all player/computers. As the piece progresses the CONDUCTOR is controlling the way in which the data is integrated, either spraying it throughout the area, (writing to every fifth, say, location) or just rolling it into the area sequentially.

The COMPUTER/PLAYERS are requested to access the data area sequentially, and to use the data recorded there for their pitch and duration decisions. When the data is used as pitch information, it should be interpreted as a sequence of standard MIDI pitch codes (middle C = 60 decimal). When used as durations, they may be freely interpreted as long as they represent a direct linear time proportion (e.g., higher numbers represent longer durations).

COMPUTER/PLAYERS are encouraged to have their pitch and duration access points NOT synchronized. The data area is 00 - FF on hub #1, and 100 - 1FF on hub #2.

COMPUTER/PLAYERS also should limit their playing to 20% of the time, distributed in any way they desire, from "1 second on, 4 seconds off" to "1 minute on, 4 minutes off." If technically possible, this limitation should be built into their program; if not, it should be faithfully followed manually.

The playing of the INSTRUMENTALIST should fall into three equal-length sections, played without a break.

In section one, he should concentrate on a small set of pitches, having a simple harmonic relation to each other, and move rather slowly in introducing new pitches. He should feel free to stop playing entirely, occasionally, and wait until he feels certain that he can hear the effect of his playing on the ensemble.

In section two, his playing should consist of more scale-like passages, again listening to try to hear the effect of changes he makes before he moves on. As the section proceeds, he should move from more pentatonic-type passages, growing out of his work in section one, to diatonic and chromatic ones.

In section three, he moves to microtonal passages and glissandi, as well as introducing other extended techniques he may find appropriate.

"Spraying" is writing the pitch data into hub memory in a non-sequential way; the pitches are preserved, but their order is not. "Rolling" preserves the pitches and their order. Note that the players must access the data in sequential order, whether or not it was written that way, and that they use the same data to generate note durations.

The pitch-to-MIDI technology of the time was not especially good at dealing with normal instrumental playing. In the only extant recording of this piece, Larry Ochs's expressive saxophone playing flummoxes the device from the outset, and it is clear that the pitch data being received by the Hub, ostensibly restricted to just a few pitches at the start, bears little relation to what Ochs is playing. The performance is lively and interesting, regardless, as the players adapt to the situation.

Every Hub composition searches for a new kind of musical structure. *Perry Mason in East Germany* is exemplary in its refusal to define what the "variables" associated with each player's activity are, leaving those associations—both incoming and outgoing—to the players' discretion. Yet somehow a structural coherence emerges, and is audible in every recording, despite not being dictated or even defined by the score.

PERRY MASON IN EAST GERMANY

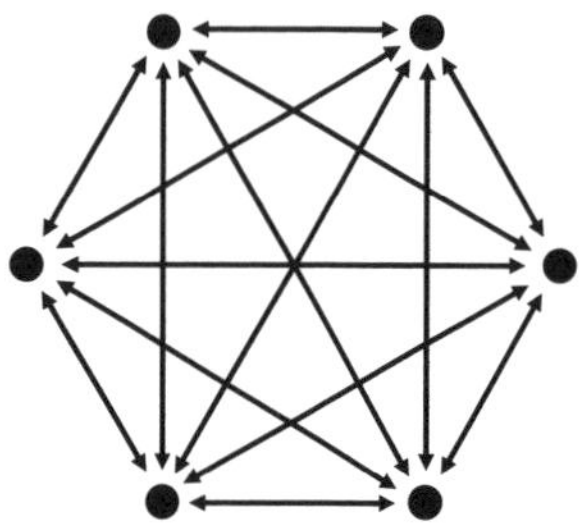

John Bischoff
September 15, 1987

Premiere: November 6, 1987 at The Lab, San Francisco

Released on *The Hub* and *Boundary Layer*

Each player runs a program of their own design which constitutes a self-sustaining musical process. Players regularly send 3 bytes to 3 other players (1 byte to each) which indicate something about the current state of their activity. Players also design their program to be influenced by 3 bytes from 3 other players (1 byte from each). To insure a balanced distribution of variables, designate senders and receivers such that all players send and receive no more—and no less—than 3 bytes.

If incoming bytes are changing too quickly for a player's musical purposes, they may be dropped as needed. If bytes are changing too slowly, players may insert random numbers that are biased in some way by the values received. The resulting web of mutual influence gives the music structure beyond any individual's planning.

HOT PIG

Chris Brown, Tim Perkis, and Phil Stone 1988

Released on *The Hub* and *Boundary Layer*

An improvisation where all players are using homebuilt electroacoustic instruments. Chris plays his Gazamba, a prepared electric piano, enhanced with homemade, electronic effects. Phil plays his "axe-thing," a guitar-like device controlling the processing and playback of sampled sounds. Tim plays "mouse guitar," a homebuilt synthesizer controller.

The improvisation was structured in this way: trio; solo (Chris); solo (Phil); duet (Phil and Chris); duet (Tim and Chris); solo (Tim); duet (Tim and Phil); trio.

THE MINISTER OF PITCH

Tim Perkis
1988

Released on *The Hub* and *Boundary Layer*

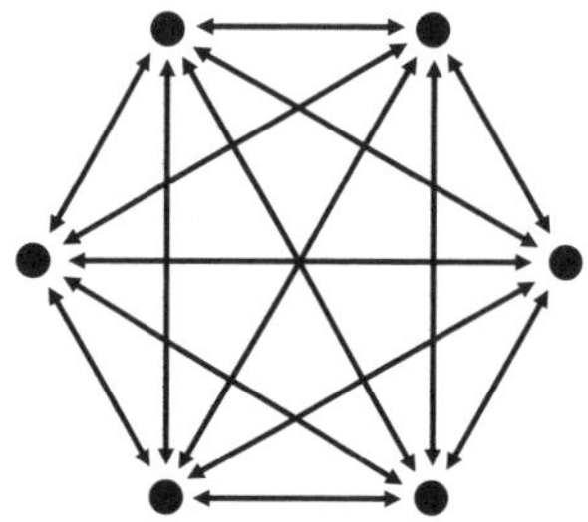

Perkis: "In most ensemble playing, the player of each instrument has some sort of responsibility for a particular aspect of the music: a bassline, keeping a rhythm, playing a main melody, etc. Using the capabilities offered when data is continuously exchanged between players, I organized responsibilities differently: one player's actions control the pitches played by all the players, another controls timing, and another is responsible for setting an overall timbre.

The Minister of Pitch was a very early Hub piece, and as such, our procedures and customs in designing new pieces was not yet nailed down. Later it became quite well-defined what it meant to dream up and propose a piece for the Hub: There was always a designer of the piece—I think without exception always one of us six members—whom we would call 'the composer,' even though this composer was *never* doing any of the things traditional composers did—that is, telling anyone else what exactly they were supposed to play! The composer would at most define the set of messages our computers would send to each other during the course of the performance of the piece, and sometimes—but not always—the meaning and intended use of these messages; how, in broad terms, we were to influence each other's playing. We insisted on thinking of ourselves as composers instead of players and, following the style of the time, playing the role of 'composer/performers,' something thought of as slightly different than a mere musician; and perhaps something else we inherited from our idols—John Cage, Morton Feldman, etc.—who seemed to model certain aspects of their practice on what visual artists do. We stand back, design this thing in front of us *on the global scale*; the machines take care of the moment-to-moment details of the music, acting on our broad general instructions, and nobody ever hits a key to make a particular note sound!

So *The Minister of Pitch,* born of this mindset, existed in a space even one step farther from hands-on instructions than later Hub pieces: It proposed not even a set of messages and actions, but rather just a class of possible musical experiments, based on an idea about ways of reassigning musical responsibilities in a new way made possible by our network connections. Our separate computers, when connected together via the hub box, in essence become one instrument (we were aware how the Javanese gamelan orchestra was considered by some to be one instrument) being played by all six of us. As such, we could therefore reassign musical responsibilities in a new and different way, giving to each individual player unique control over some global aspect of the music: say, I decide what pitches everyone plays, John decides the rhythms or precise times each player plays, Chris decides which of a set of timbres are used next by each player, and so on. In essence we were trying to make a piece that serializes a set of musical parameters, and assigns responsibility for them severally to different players.

But that general concept was as far as it went: whereas In later, more 'normal' Hub pieces the composer would set up exact messages to be sent to implement the composition's scheme, in *The Minister of Pitch* it was left for each Minister to decide *how* to specify his own domain. If in a certain performance Phil is assigned to be the Minister of Pitch, he might give players a palette of pitches they could choose from that would change over time, or tell someone more specifically that their next note should be F#, or specify a currently active pitch range for each player—whatever scheme Phil wants. Likewise, in a particular performance, if Scot was the Minister of Timbre, he would come up with his own rules to communicate throughout the piece. The schemes would change at different performances, but of course in practice they tended to settle down to precise reusable rules.

This was charming in a way, but perhaps too open, too confusing, and—especially annoying when we've reached this point in the band's history—too undocumentable.

In fact, I may be mis-remembering completely what we actually did."

WHACKERS

Scot Gresham-Lancaster

October 15, 1988

Released on *The Hub* and *Boundary Layer*

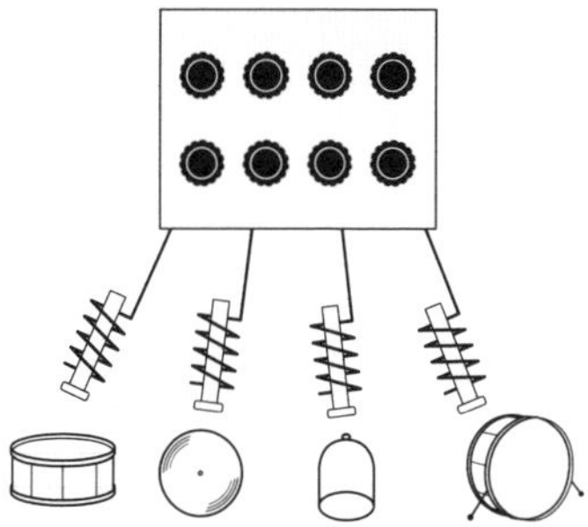

Gresham-Lancaster: "The whackers were computer-controlled solenoids on wood clamps that could strike objects they are next to. Tempo is a variable and played so there is a consistent rhythm. A range of random choices and amplitudes were generated by a Forth program played in real time by changing ranges. The sound was the result of the objects that are struck. The output of an 8-bit parallel port is multiplexed to four sets of 8, so 32 drivers for an industrial stepper motor driver, so there were 32 whackers.

In 1984 I had the rare opportunity of working closely with Iannis Xenakis during his residency at Mills College when I was the Technical Director there. He was a real stickler for all aspects of sound reproduction. At one point he walked up to one of the speakers that we were EQing and said 'the reliance on speakers is the most serious problem with electroacoustic music.' This idea of the weakness of the core instrument of all computer music stuck with me and was the inspiration to create the whackers, a percussion robot that was also inspired by my collaboration with Trimpin that same year. The aspect of this piece that does not come across in the recording is the dimensionality of the listening experience. The sound the whackers make are physically in the space itself and not the result of panning or mixing between fixed location loudspeakers to fool the ear with an imaginary space."

DOVETAIL

John Bischoff
April 14, 1989

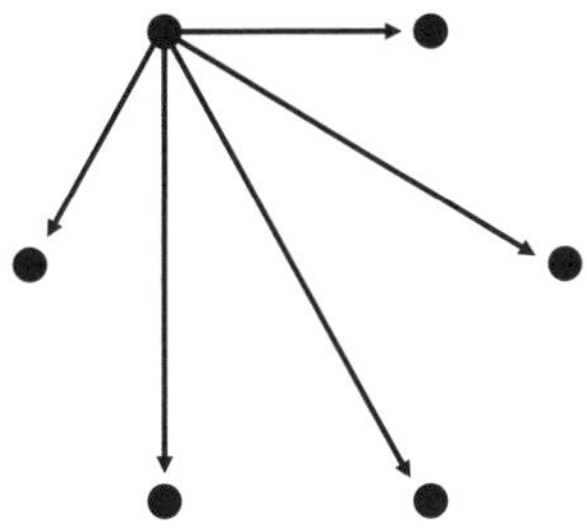

The same rhythmic processes are also used in Bischoff's *Action at a Distance.*

Perkis: "As in *The Minister of Pitch*, musical responsibility is apportioned in an unconventional way. John provides a rhythmic process which is being computed live and transmitted to the other players. The other players search for patterns in this input and play phrases determined by this analysis."

Bischoff: "The rhythmic process transmitted to the other players consists of eight real-time fractal function generators which create a counterpoint of semi-independent rhythmic sequences. The ascending, descending, and stationary conditions of the functions determine their rhythmic properties. Each time a function enters a stationary state, it automatically alters the tempo of its neighbor. A global tempo control operated by the composer biases the overall tempo faster or slower. The group effect created by two players deriving their actions from a third—a kind of variable shadowing—evokes a complex chorusing quality. One hears both an overall coordinated flow—movement in a particular direction—and also detailed departures and deviations that sometimes linger."

HUB RENGA

The Hub, with Ramon Sender Barayon, and poets from the WELL

September 7, 1989

Released on *Wreckin' Ball*

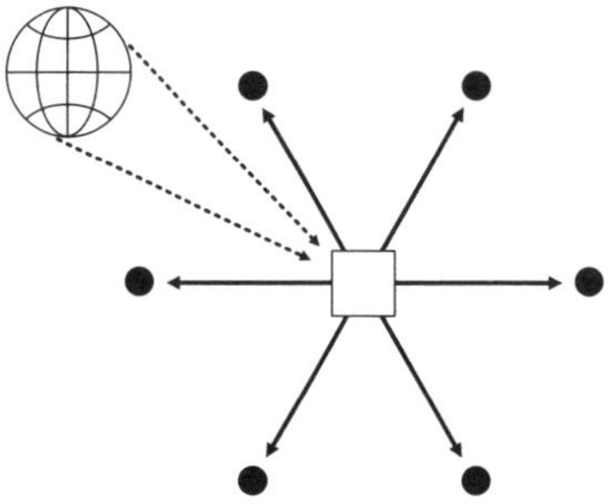

This ambitious piece is fully in the spirit of Bay Area techno-utopianism. Ramon Sender, the instigator, was one of the founders of the San Francisco Tape Music Center in 1962, a long-time communard at Morningstar Ranch, and an early participant on the WELL, one of the first Internet online communities.

Gresham-Lancaster: "*Hub Renga* epitomized the openness of the community of artists that was emerging in the post-hippie techno-utopia of the Bay Area, With the network open to all comers and the technology simplified, we assumed it would be an egalitarian victory for art. The varying range of taste and innate talent made for a pastiche that lacked finesse and cohesion, and despite the best intentions of the contributors, the results were mixed. I think with curation and rehearsal, the sort of unique structure put forward in *Hub Renga* is very promising and should be revisited and refined. I was always fond of this piece as it represented a deep integration of protocol-social media, computer network music, and community collaboration that to my knowledge has not been matched."[3]

3 Scot Gresham-Lancaster, "The Aesthetics and History of the Hub: The Effects of Changing Technology on Network Computer Music," *Leonardo Music Journal* 8 (1998): 39–44.

No formal spec for the piece was made, as it developed in a series of collaborative emails and online interactions and culminated in a live 90-minute interactive performance broadcast on Berkeley radio station KPFA. Tim Perkis's notes describe it well:

"Renga is a traditional Japanese poetic form in which different poets contribute lines, continuing from the previous poet's work. In this piece, the Hub composers programmed their systems to respond to certain pre-determined 'power words' in a text stream coming from participating writers in the poetry conference of the WELL, an online community. In live performance, the WELL poets could, from their homes, type in their poetic line and hear it immediately read aloud on the air, while it simultaneously and automatically influenced the evolution of the Hub's music. It's hard to describe the peculiar quality of community feeling that arose in the project: There was something authentically new about people being able to interact socially, while at the same time doing the traditionally more private work of writing and composing music."

In the weeks before the broadcast, 64 "power words" were suggested and collected in the poetry conference of the WELL, and then shared with the members of the Hub, each of whom designed their own sonic response to each. During the performance, renga using these words were composed in real time by anyone who cared to participate on the WELL, and sent on to the Hub. These were the "power words":

EMBRACE ECHO TWIST RUMBLE
KEYSTONE WHISPER CHARM MAGIC
WORTH KAISER SCHLEP HABIT
MIRTH SWAP SPLIT JOIN PLUS
MINUS GRACE CHANGE GROPE
SKIP VIRTUOSO ROOT BIND ZING
WOW EARTH INTIMIDATE OUTSIDE
PHRASE HONOR SILT DUST
SCAN COFFEE VERTIGO ONLINE
TRANSFER HOLD MESSAGE
QUOTE SHIMMER SWELL RICOCHET
POUR RIPPLE REBOUND DUCK
DINK SCINTILLATE OLD RETREAT
NON-CONFORMIST FLOWER SKY
CAGE SYNTHESIS SILENCE CRUMP
TRUMP IMMEDIATE SMACK BLINK

The Hub, 1989, (L–R) Chris Brown, Scot Gresham-Lancaster, Mark Trayle, Tim Perkis, Phil Stone, John Bischoff. Photo: Jim Block.

Bischoff's earlier piece *Dovetail* also uses eight simultaneous rhythmic processes as a framework for players to riff on. Like *Dovetail, Action at a Distance* is based on his solo piece *The Curve Behind the Line.* Bischoff sends the control data from his solo performance to the rest of the Hub for interpretation and elaboration.

ACTION AT A DISTANCE

John Bischoff
September 22, 1989

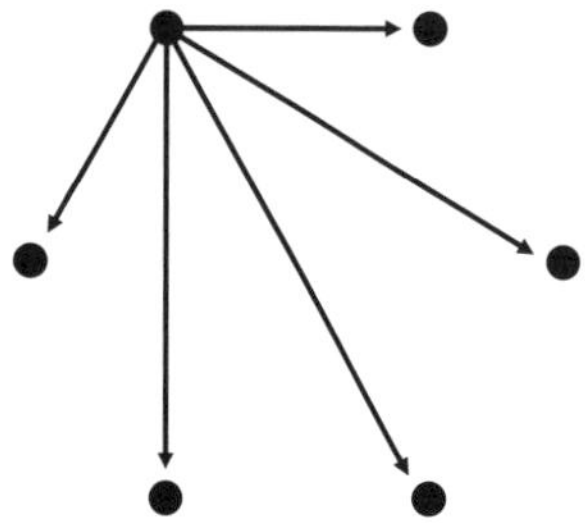

One player designs an algorithm that generates six independent rhythmic parts as a framework for all players, including themselves. Each part must consist of rhythmic phrases 3 to 8 seconds long that alternate between regular and irregular patterns. All patterns within a part should be based on a fixed ictus established for that part. The ictus rate for each part must be different from all other parts and at the same time related by whole number multiple.

All rhythmic events generated by the algorithm are sent to the Hub using a `trigger` message with an integer `(1-6)` where the variable indicates which one of the parts the rhythmic trigger is from.

Each part serves as a template from which player sounds may be triggered or independent musical processes begun. Players are free to choose which part among the six they respond to. They also may ignore triggers or make additional sounds independently at any time. All other sound parameters are open to player choice.

HUB 2.0

The next hub was a commercially-made device, the Opcode Studio 5 MIDI interface. Scot Gresham-Lancaster adapted it to the Hub's purposes in 1990. It was a much faster device and the communications protocol pushed messages in real time, so that performers could parse them immediately. This led to a new repertoire of pieces.

During this period, MIDI-controlled synthesizers were common, and many members of the Hub relied on them as sound sources. The MIDI protocol was designed to enable control of synthesizers by computers or remote keyboards, so its primary communication is in the form of Note On and Note Off messages. Even though the Hub used the MIDI protocol in unconventional ways, the synthesizers run by it were note-based, and this influenced to some degree the pieces written in this period.

Around 1994, the World Wide Web became active, and the word "network" gathered new meanings. The ARPANET protocols on which it was based had been built to be robust, scalable, and reliable, but not especially secure. As the Web became a commercial space in the 1990s, security became a new and pressing concern. Of this evolution, Internet historian Janet Abbate says: "They thought they were building a classroom, and it turned into a bank."[4]

For the Hub, this period culminated in the 1997 "Points of Presence," a network specification to permit the group to perform together from three separate, remote locations over the Internet. The experiment was a technical and aesthetic failure, and led to a hiatus. Even though the Hub was the first band to perform "telematically," in 1987, then as now they found that physical separation was antithetical to the spirit of the band. Technically, Gresham-Lancaster notes: "difficulties tunneling messages through the increasingly rigorous security of the three campuses where we were playing made a recipe for disaster."[5] It's difficult not to read this as a culture clash—network "security" is in clear opposition to the Hub's purpose of sharing data, while restrictions on sharing are essential to commerce, the purpose to which the Web was increasingly dedicated.

4 Craig Timberg, "A Flaw in the Design," *Washington Post*, May 30, 2015.

5 Scot Gresham-Lancaster, "A Personal Reminiscence on the Roots of Computer Network Music," *Leonardo Music Journal* 27 (2017): 71–77.

WAXLIPS

Tim Perkis
1991

Released on *Wreckin' Ball* and *Boundary Layer*

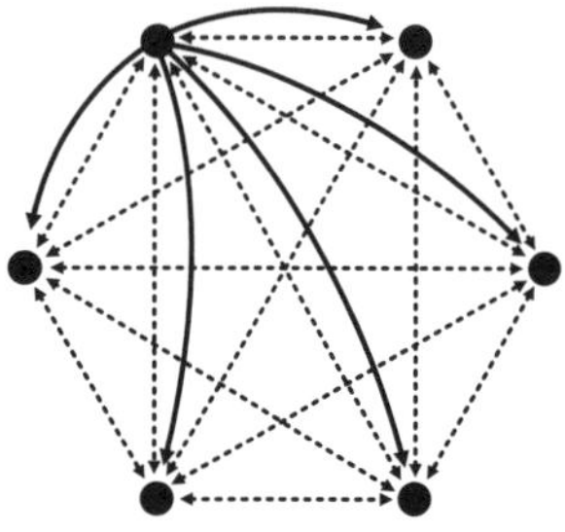

Ok, hubbies this is a score. I'd like to try to do one where you can really see if there is any emergent pattern to a static setup, where each station acts in a fixed, predictable way, but the interconnects are so complex that the overall behavior is still groovy.

The piece is simple. Each player does essentially the same thing: take Note On MIDI messages in, transform them in a regular way that I'll specify below, play the new transformed note and send out a copy of it to somebody. I'll "seed" the process at the beginning of the piece or section by sending out a few notes to start.

The transformation can be anything you want, within these limitations:

→ One note in, one note out. For every possible MIDI note and channel input combination (`127*5` or `635` total) you define a unique transform to some other MIDI note and channel combination. Within any one performance of the piece this mapping is fixed: Each time a particular note
on a particular channel is received the same transformed output is sent.
No random number changes, knob or slider or button adjustments, no algorithms which depend on previous states of your machine or previous input. A simple, fixed mapping.

For example, the mapping could be as simple as: send the same note out, transposed up a fifth, to the next channel `(mod 5)`. Or it could be as complex and arbitrary as a randomly generated (beforehand!) lookup table giving transforms for each of the 635 cases.

→ These mappings are fixed for the performance of any one section, but you are encouraged to provide yourself with controls which allow you to easily define new mappings for new performances of the piece. I really don't know what this will sound like, but I imagine a concert performance of the piece would involve playing several sections, stopping each when it gets boring, changing the mappings and starting over.

→ If you wish you can incorporate a delay into your mapping— that is, you can define for each input case not only an output note and channel, but also an amount of time you wait before playing your note and sending your output. This delay should be a fixed one-to-one mapping, just like the note and channel mapping, not subject to any user, algorithmic or random adjustment within any one section.

→ You may also transform the velocity of the note you receive, again, only through a simple mapping, and with the additional restriction that you only transform the velocity by the amount `-1, 0, or +1`. For example, you might decrement velocity for all notes below middle C, and increment all others. Or decrement velocity on notes received on odd channels, increment velocity on notes received on evens, except for notes you get from Scot, which you pass with velocity unchanged. You get the idea. Any mapping you want, but fixed and one-to-one.

→ Avoid droning voices: no Note Off messages are being sent around, and you should set your synthesizer to percussive type sounds which die by themselves without Note Offs or generate your own Note Offs on your local MIDI chain.

Another Bischoff solo piece adapted for Hub use. Each player's computer monitors that player's sound, and thereby tells the others how often to transition their sounds (based on their own volume), and how fast the transition should be (based on their own pitch, as determined by the player). The only performance action available to each player is to adjust their own volume, and optionally to change the sounds they are using. As the data from each player's amplitude and pitch propagates through the network, the following tendencies emerge:

fast triggers + fast speeds = rapidly changing textures

fast triggers + slow speeds = texture hardly changes at all

slow triggers + fast speeds = quick transitions to each new texture

slow triggers + slow speeds = continuous gradual change between textures

THE GLASS HAND

John Bischoff
November 19, 1991

Released on *Wreckin' Ball* and *Boundary Layer*

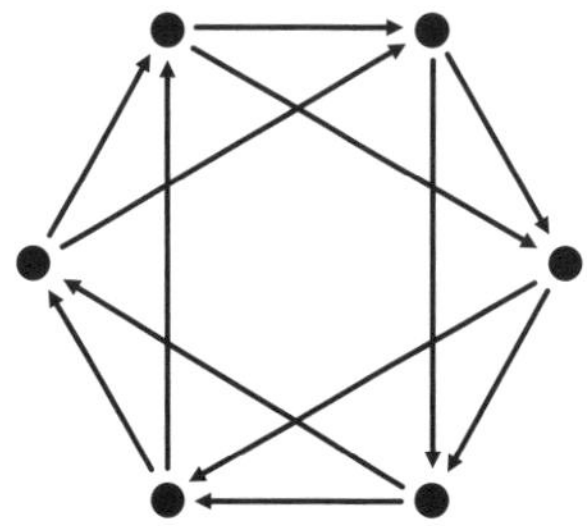

This piece involves multiple sonic layers that transform themselves at differing rates. Each player generates one layer or ribbon of sound that can be thought of as an evolving texture. The texture must be able to evolve slowly or rapidly, or at any rate in between. The rate of evolution is determined by data interactions among the players. It is easiest to think of the textures as having steady states—or presets—which are migrated between to create the evolutionary flow. Each player may define these steady states ahead of time and then migrate between them in performance. Adjust your volume level within the overall group mix as the piece unfolds.

MESSAGES

→ `Trigger` messages start a migration toward a new texture. Send `trigger` messages at a rate proportional to your current volume level (full volume = 10 triggers per second; silence = 1 trigger every 10 seconds).

→ `Speed` messages set the migration rate from 0.1 to 10 seconds. Send `speed` values proportional to your overall pitch tessitura (lowest pitch = migration rate of 10 seconds; highest pitch = migration rate of 0.1 seconds).

→ Messages are sent in a ring—player 1 sends `triggers` to `player 2` and `speeds` to `player 3`; player 2 sends `triggers` to player 3 and `speeds` to player 4; etc.

CRYBABY

Mark Trayle
1991

Released on *Wreckin' Ball*

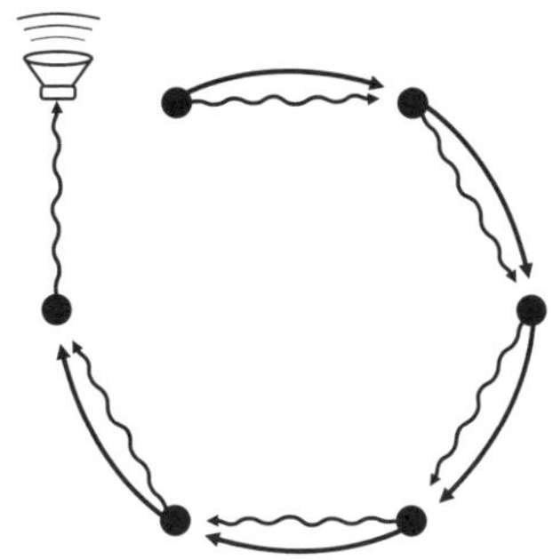

In performance, Trayle provided all the audio (usually samples of guitars and monster trucks), which went through a daisy-chain of processors. The data flow in the network is also a daisy-chain.

This piece was inspired by the Crybaby wah-wah pedal.

THE GENERAL IDEA
The audio from some sound source goes through a chain of signal processors. The output of each signal processor goes to a mixer input, then the whole mess is mixed and diffused in hi-fi stereo. various parameters of the signal processors are controlled by data passed around through the Hub `/Opcode` box. The data travels in a loop. `Audio:chain /Data:loop`.

I know that John, Scott, Chris, and I have signal processors… I'm not sure about Phil and Tim. Whoever doesn't have a signal processor can play/be the sound source. Right now I'm thinking CD player for sound source… maybe a sound-fx CD. Or maybe a radio. Tim? Shortwave?

SPECIFICS
The thing about the wah-wah pedal is that you can't get from A to C without going through B; it ain't discrete (in more ways than one).

So the processor players need to select an effect that already has an oscillating component (I'll probably use a `chorus+delay` or `flange+delay`, for example), or write some code to make some portion of a chosen effect oscillate. The data from one of your fellow Hubsters will determine the frequency of this oscillation. Still with me? Okay, data only flows one way through this loop. You get your data from the guy "upstream" from you. That data will come in the form of a MIDI Note On, where `note value = frequency of oscillating effect` and `velocity = width (amplitude) of the effect`.

Once you use the data, change the value of the Note On and then pass it on to the guy downstream. Rule for changing Note On: use a translation table, i.e., no adding of random numbers to Note Ons, no xoring it with your favorite prime number.

WHEELIES

Chris Brown

January 19, 1992

Released on *Wreckin' Ball* and *Boundary Layer*

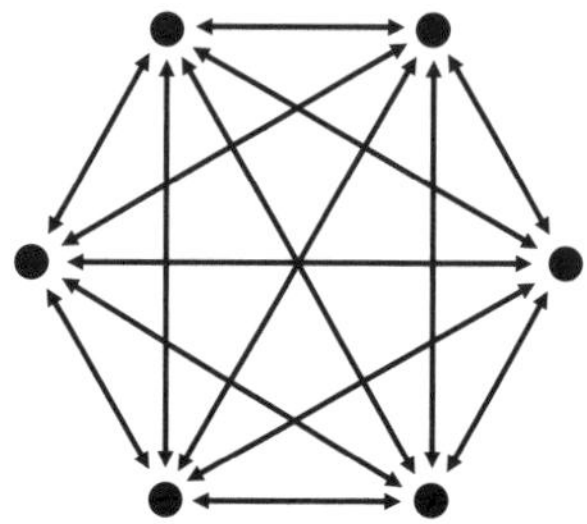

This drum-machine-inspired piece defines an interactive behavior in which networked players can change the rhythmic character of each other's performance. All players share a common pulse generated by one player which is variably interpreted by their programs to enable complex polyrhythmic relationships. They can directly alter each other's interpretation of the pulse by sending messages.

INSTRUCTIONS

One player-conductor generates and broadcasts `Timing Clock` messages (`$F8`) to all players. The conductor also controls tempo, which can change at any time, and all players lock and count incoming clocks, playing only in sync with them. Players choose their own sounds and change timbres on their instruments freely. Percussive sounds or samples are preferred. All players both generate and receive MIDI Start (`$FA`), Continue (`$FB`), and Stop (`$FC`) messages. They respond to them as follows:

[1] Start → reset your Timing Clock counter to zero and send three rhythm parameter messages to any other players in the group. Use GUI controls while the music plays to select the data that will be sent automatically when the next `Start` message is received. These specify the player to be addressed, the rhythm parameter to be controlled, and a data integer in the range of 1–10. The rhythm parameters are:

Ictus → number of Timing Clocks of the shortest duration between played notes;

Density → multiplier of Ictus to increase the duration of beats and cycles of beats;

Meter → number of Ictus * Density beats in a cycle of repeating timbres or notes to play.

Begin the piece with your own default values for these parameters. But changes received from other players during the course of the piece must be obeyed and may not be changed by you during the course of the piece!

[2] Stop → stop playing sound on your own system, but don't stop counting Timing Clocks and don't reset your counter. Watch for a `Start` or `Continue` message to follow—this functions as a group mute.

[3] Continue → resume playing sound on your own system with current counter values intact.

Brown: "*Wheelies* used the computer network band as a giant, multi-limbed, rhythm machine. The purpose was to create variations on complex rhythmic synchronizations among the players' machines, while allowing players to control each other's rhythmic performance. Each player programmed the implementation of their system and could change the voicing of their instrument's rhythms, but all were locked into a tempo set globally for the group, while their Ictus, Meter, and Density parameters were set and changed during the performance by other players, but never by themselves."

Bischoff and Brown: "Characterized not just by its spunky rhythmic character, the multi-dimensionality of the networking protocol in *Wheelies* was unique. While Chris controlled the global tempo, dominating that aspect of the conductor's role, any player could start or stop the whole group and thus conduct the phrasing of its rhythms. And an unusual kind of ensemble behavior resulted from the rule that players could only change each other's rhythmic behavior and not their own. One listened to recognize which part of the whole group sound each player was creating, and then tried to shape it by specifying parameter changes for them. Everyone could affect everyone else's sound, but no one could control everything about even their own sound. Because data exchanges happened only at Start signals, these were like moments when the cards were shuffled for a hand which would proceed unchanged in character until the next Stop. The tempo and rhythm of this rhythmic change was under the control of any player, all the time, which meant that the whole group became attuned to shared control of this most important structural element in the piece."[6]

6 John Bischoff and Chris Brown, Ibid.

LISTENER'S DIGEST

Scot Gresham-Lancaster

1992

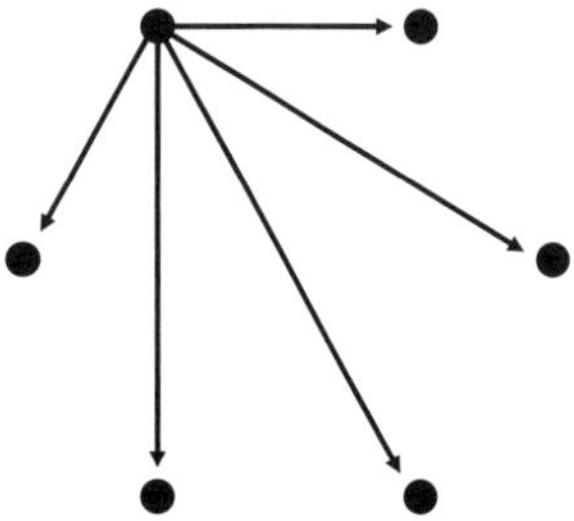

Each of the six players is responsible for a monophonic melody that is created by consensus with all the other five voices. The choices of pitch will be based on decisions made in accordance with several parameters. The choice of each pitch will be entirely context dependent and will arise out of the context of the other's choices.

I will be passing MIDI Note On messages to each player that will inform about what the current `1:1` is for their melody and what the peak dynamic (e.g., velocity) is. Zero velocity message means rest, anything else means playing is at your discretion These `note` messages will be passed in a rhythmic manner and if possible players should try to perform pitches in conjunction with the notes as they arrive.

Below are tables of criteria that are based on information from Barlow's "Two Essays on Theory" from the *CMJ* which will provide an easy table-driven base from which to base your melody generating algorithms. If possible, at first, always go for the interval with the greatest "harmonicity value" in the context of the notes being played by other individuals. Every time you play a note, broadcast your note choice to each of the other players. For example, you communicate a note choice of the `3:2` of the current intonation as `159` (MIDI Note On `to ch. 16`) `03 02`; for `27:16` it would be `159 27 16`. When you play a new note make sure and turn off the old one (e.g., `159 03 00` and `159 27 00`). Each player will be responsible for keeping track of which intervals are available. The cumulative "harmonicity" of the current vertical sonority will be sent as a running 14-bit Pitch Bend on `channel 16`. So in this case if you get a `Note On` message, you are dealing with a `3:2` and `27:16` from the table below; this equals about `.35666` or `$.8562`. The higher the value the better.

Making a note choice should be done in whole or half steps in the opposite direction of the last leap. Leaps should be arbitrarily executed every once in a while, when no scalar choice is available or possibly when the edge of the tessitura is reached. Since I am asking for a monophonic melody, that means you can use MIDI Pitch Bend to sharpen any equal-temperament pitch to exactly match the frequency needed. I will be adding MIDI note number plus 14-bit Pitch Bend value to the Harmonicity table as soon as I get finished

calculating all of those. I will assume that you have set the global parameter on your MIDI sound producing unit to the range of whole tone in either direction (the standard default). For those of you using Amiga local sound, simply multiply the frequency of the `1:1` times the `Frequency Ratio` in the table below to get the frequency.

The overall structure of the piece will start off as a well-tuned chorale, but as this goes along, vagrant "free style" intonation will be introduced to some of the players information, (i.e., differing `1:1`). As this begins to happen more raucous timbres should be employed. The rhythmic activity level is canonic in nature and will increase in frenzy as the piece progresses; in keeping with this the melodic choice will become more free and less reasoned as it progresses. At some point the activity will begin to slacken and we will wind down back to the tame chorale of the opening section as we hunt for an arbitrary cadence.

HARMONICITY TABLE

Interval	Frequency	Harmonicity
0	1:1	2.000000
111.731	16:15	0.076531
182.404	10:9	0.078534
203.910	9:8	0.120000
231.174	8:7	0.075269
266.871	7:6	0.071672
294.135	32:27	0.076923
315.641	6:5	0.099338
386.314	5:4	0.119048
407.820	81:64	0.060000
435.084	9:7	0.064024
498.045	4:3	0.214286
519.551	27:20	0.060976
701.955	3:2	0.272727
764.916	14:9	0.060172
813.686	8:5	0.106383
884.359	5:3	0.110294
905.865	27:16	0.083333
933.129	12:7	0.066879
968.826	7:4	0.081395
996.090	16:9	0.107143
1017.596	9:5	0.085227
1088.269	15:8	0.082873
1200.00	2:1	1.000000

Approximate tessituras assigned in terms of equal-temperament pitch and MIDI note number:

John	b2 - b4	(59 - 83)
Tim	g2 - g4	(55 - 79)
Chris	e2 - e4	(52 - 76)
Scot	a1 - a3	(45 - 69)
Mark	f1 - f3	(41 - 65)
Phil	d1 - d3	(38 - 62)

As a postscript I should add that I am intending to make a change to the Studio 5 patch so that we all have a global channel of MIDI channel 16 to communicate big info, like the Pitch Bend `/group harmonicity number`.

Scot email to hubbies before first performance: "I hope I am not asking too much… here is the long awaited table for 'Listner's Die Jest.' I have given lots of new information here that should make the task at hand much easier. The last column is the one that you can base all your harmonic decisions on in a much more palatable form (i.e., unison is `number 1`**, an octave is** `number 2`**, a fifth is** `number 3`**, a fourth is** `number 4`**, etc.). This is much easier than the harmonicity factor before. To play a MIDI note in tune using Pitch Bend, take the** `1:1` **I give you as a Note On, add the offset, then use one of the Pitch**

Bend offsets in the table. Some systems parse the 14-bit number into two 7-bit MIDI data bytes and others need two separate 7-bit bytes after the bend command. I have assumed that the `LSB` is `00`. You can tweak that if you think these intervals are out of tune, but when I tested it these were within a cent. Remember I want to start with harmonically simple sustained tones at first and move into harmonic complexity as the piece accelerates."

ratio		cents	14^{bit} hex	14^{bit} dec	7^{bit} hex	7^{bit} dec	harmonic priority
1:1	0	0	2000	8192	40	64	1
16:15	0	111.7	3100	12544	62	98	17
10:9	1	182.4	2C00	11264	58	88	15
9:8	1	203.9	3000	12288	60	96	5
8:7	2	231.2	2480	9344	49	73	18
7:6	2	266.9	2780	10112	4F	79	19
32:27	2	294.1	2E80	11904	5D	93	16
6:5	3	315.6	2380	9088	47	71	10
5:4	4	386.3	1D80	7552	38	59	6
81:64	4	407.8	2100	8448	42	66	24
9:7	4	435.1	2500	9472	4A	74	21
4:3	5	498	1F80	8064	3F	63	4
27:20	5	519.5	2180	8576	43	67	22
3:2	7	702	2004	8196	40	64	3
14:9	7	764.9	2780	10112	4F	79	23
8:5	8	813.7	2280	8832	45	69	9
5:3	8	884.4	2000	11392	59	89	7
27:16	9	905.9	2100	8448	42	66	12
12:7	9	933.1	2580	9600	4B	75	20
7:4	9	968.8	2800	11008	56	86	14
16:9	10	996.1	1F00	7436	3E	62	8
9:5	10	1017.6	2200	8704	44	68	11
15:8	11	1088.3	1D80	7552	3B	59	13
2:1	12	1200	2000	8192	40	64	2

A detailed and uncharacteristically complicated spec for this composer. The piece was inspired by an article in *Computer Music Journal* by the composer Clarence Barlow suggesting a numerical measure for the harmonic complexity of musical intervals, which Barlow termed "digestibility."[7] (Other such measures exist, such as Leonhard Euler's *gradus suavitatis*.) It continues the Hub's long involvement with just intonation and alternative tunings (see *Tudor, Farabi, Hertz Donut, Lip Bomb, Lou Drift*). Most of the complexity stems from the technical difficulty of playing non-tempered pitches on MIDI synthesizers. To play any pitch outside of 12-tone equal temperament, the correct amount of "pitch-bend" offset retuning must be sent along with the MIDI Note On command. The composer's painstaking work in calculating all the offsets in hexadecimal and decimal, documented here, is a testament to similar "under-the-hood" work done by all the Hub composers on nearly every piece.

"Free style" intonation is an innovation by composer Lou Harrison described in more detail in the notes to *Lou Drift*. By the time of *Lou Drift*, the Hub was using software synthesizers that made these MIDI-based calculations unnecessary.

7 Clarence Barlow, "Two Essays on Theory," *Computer Music Journal* 11.1 (1987), 44–60.

		Cents	14bit Hex	14bit Decimal	7bit MSB HEX	7 bit MSB Decimal	Harmonic Priority
1:1	0	0	2000	8192	40	64	1
16:15	0	111.7	3100	12544	62	98	17
10:9	1	182.4	2C00	11264	58	88	15
9:8	1	203.9	3000	12208	60	96	5
8:7	2	231.2	2480	9344	49	73	18
7:6	2	266.9	2780	10112	4F	79	19
32:27	2	294.1	2E80	11904	5D	93	16
6:5	3	315.6	2380	9088	47	71	10
5:4	4	386.3	1D80	7552	3B	59	6
81:64	4	407.8	2100	8448	42	66	24
9:7	4	435.1	2500	9472	4A	74	21
4:3	5	498.0	1F80	8064	3F	63	4
27:20	5	519.5	2180	8576	43	67	22
3:2	7	702	2004	8196	40 04	64 04	3
14:9	7	764.9	2780	10112	4F	79	23
8:5	8	813.7	2280	8832	45	69	9
5:3	8	884.4	2D00	11392	59	89	7
27:16	9	905.9	2100	848	42	66	12
12:7	9	933.1	2580	9600	4B	75	20
7:4	9	968.8	2B00	11008	56	86	14
16:9	10	996.1	1F00	7936	3E	62	8
9:5	10	1017.6	2200	8704	44	68	11
15:8	11	1088.3	1D80	7552	3B	59	13
2:1	12	1200	2000	8192	40	64	2

1 = 1:1 A –
2 = 2:1 A –
3 = 3:2 E –
4 = 4:3 D ↓
5 = 9:8 B
6 = 5:4 D♭↓
7 = 5:3 G♭↓
8 = 16:9 G↓
9 = 8:5 F↑
10 = 6:5 C↑
11 = 9:5 G↑
12 = 27:16 F♯↑
13 = 15:8 G♯↓
14 = 7:4 F♯↑
15 = 10:9 B♭↑
16 = 32:27 C↓
17 = 16:15 B♭↑
18 = 8:7 B↑
19 = 7:6 B↑
20 = 12:7 F♯↑
21 = 9:7 C♯↑
22 = 27:20 E♭↑
23 = 14:9 E↑
24 = 81:64 C♯

7 =

EVERET VERBUM

Alvin Curran
1993

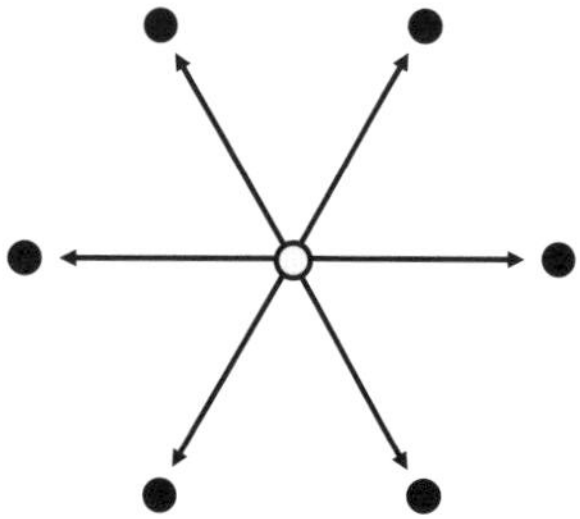

Everet Verbum is derived from the "Delta" section of *Erat Verbum,* a 6-part sound work commissioned by the Studio Akustische Kunst of the WDR and still in progress. Here, sections of John Cage's illustrious Charles Eliot Norton Lectures from 1988–89 (later published as *I–VI* by the Harvard University Press) are read by Cage, himself, and fed to the Hub for perusal and instant retranslation into Morse code; the resultant dot and dash fantasy was mixed live by Curran on his way to the Hub concert.

Everet Verbum

is derived from the "Delta" section of ERAT VERBUM a 6 part sound work commissioned by the Studio Akustischen Kunst of the WDR, and still in progress. Here sections of John Cage's illustrious Norton Lectures (on "I–VI") 1989 read by Cage are fed to Hub for perusal and instant retranslation into Morse Code – the resultant dot and dash fantasy is mixed live by Curran on his way to the Hub Concert.

Curran's original manuscript for *Everet Verbum.*

WAXLIPS II

April 23, 1993

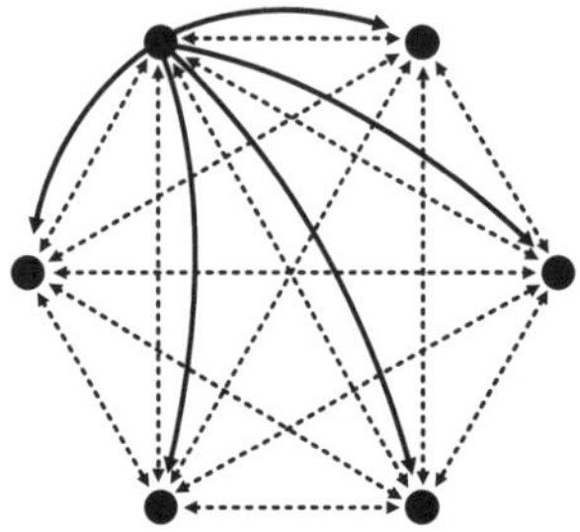

Addition to *Waxlips* for Berlin, Moers and ROVA 1993.

In the new version, an additional structure is added which defines a set of pitches which are legal for use in defining the mapping table. Using the Hub blob data standard, addresses `0-11` in the `ch 12 kd blob` will hold pitch class designators in `range 0-11` where `0 = C`, `1 = C#`, etc. This table should be continuously updated, but it shouldn't change the network behavior until a new table command is sent.

Around this time, the Hub started using the Opcode Systems Studio 5 as their MIDI-based data-sharing interface. This piece directly exploits the new capabilities of the device—its higher speed and addressable "push" messaging—and incidentally uses MIDI Note On messages, although these could be used to trigger any sound a player chose.

Waxlips was often used to debug the new network in rehearsals and before performing. Phil Stone writes: "It became the de-facto 'shakedown' piece for Hub tech rehearsals; if there were *any* 'holes' in a given network setup, from either faulty hardware, buggy software, or incorrect connection, the piece would not work correctly."[8]

In performance, whenever the network behavior became uninteresting, Perkis started a new section by sending a command to all players to change their mapping tables, then sprayed a few Note On messages to restart the process. *Waxlips II* included a pitch-class table to be used by all players; along with the player-specified mappings, this table was updated by Perkis whenever he started a new section.

8 Phil Stone, "Non-Mathematical Musings on Information Theory and Networked Musical Practice," *Organised Sound*, in press.

VEX

Scot Gresham-Lancaster

May 1, 1993

Released on *Wreckin' Ball* and *Boundary Layer*

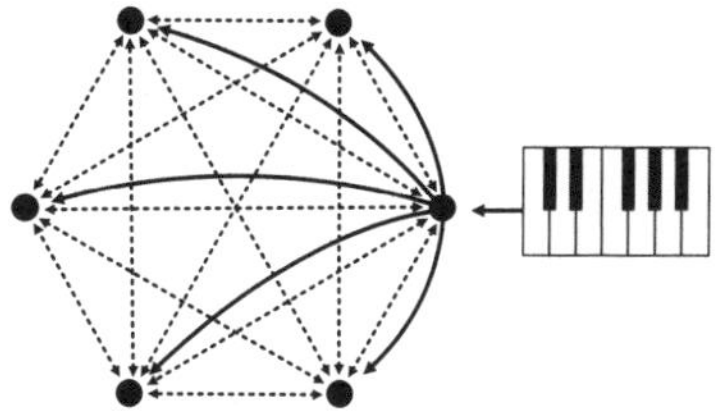

The simple instructions are: when you receive a note transpose it up or down a major or minor third while decreasing the associated velocity message of that note by 10. After playing that note for its full duration pass it on to two other players.

The title comes from Eric Satie's *Vexations*. The composer would feed the notes of *Vexations* out to the Hub, one player at a time. In the MIDI spec, note "velocity" is associated with volume and has a maximum value of 127. The rule decreasing each subsequent velocity by 10 guarantees that a note will "die" after a maximum of 12 diminishing repetitions. Thus the overall shape of the piece is many overlapping decrescendos. The composer notes: "This ended up giving a musical feeling of a nocturne as the chorale was always surrounded with complex triadic structures diminishing away from the original material."

Vex was sometimes performed as a collaboration with composer/pianist Alvin Curran and the ROVA Saxophone Quartet, with the acoustic instruments playing the Satie chorale simultaneously with the Hub's processing.

Curran improvised on a Yamaha Disklavier piano, and the MIDI output from that instrument was broadcast to all the players to use as they wished. In performance, this piece would often precede, and segue into, *Vex*. Gresham-Lancaster writes: "Alvin and I worked out a distinct musical cue for when I would start to play the chorale that fed the Hub data for *Vex* instead of his live Disklavier playing."

ELECTRIC RAGS III

Alvin Curran
May 19, 1993

Released on *Wreckin' Ball*

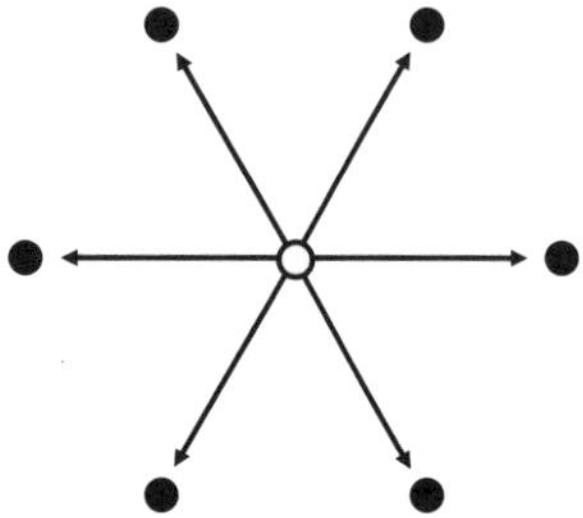

"I sit and play their stories. She sits and hubs it. We sit and talk; codes and pianos are barely mentioned—all on the verge of innuendo until she drops the bag of groceries. That triggers events parallel to those in Sarajevo but not unlike those in Fanny Mendelsohn's Sunday salons a century ago—the forte pianoforte—sonic citadel of the West leads the way into the electronic bush where aboriginal cries are translated into Sea-Sound. Doe a deer. Re a rat. "B" a was. Fig an african fruit. Imitate, ignore, dominate, tacit, and be free—this is how we play and how we hub."

E-Rags III for + with The HUB

I SIT AND play their stories. She sits and hubs it. We sit and talk; codes and pianos are barely mentioned – all on the verge of innuendo until she drops the bag of groceries. That triggers events parallel to those in ~~[illegible]~~ Sarajevo ~~and~~ but not unlike those in Fanny Mendelsohn's sunday salons a century ago – the forte piano forte – Sonic citadel of the WEST leads the way into the electric bush where aboriginal cries are translated into Sea-Sound. Doe a deer Re a rat. 'B' a was. Fis an african fruit. Imitate, Ignore, Dominate, Tacit, and be free – this is how we play and how we hub.

19.V.93

Curran's original manuscript for *Electric Rags III.*

Another exploration of just intonation, requiring Stone to calculate MIDI Pitch Bends, as Gresham-Lancaster did in *Listener's Digest.* Note Stone's wry comment, characteristic of the Hub's questioning of musical norms: "We will have to make sure we are in tune. I am deeply sorry about this."

Stone recalls: "The Hub performed *Hertz Donut* at Georgia Tech in front of a large screen projecting moments from Sergei Eisenstein's *Battleship Potemkin,* which it accompanied rather well!"

HERTZ DONUT

Phil Stone
1994

Released on *Boundary Layer*

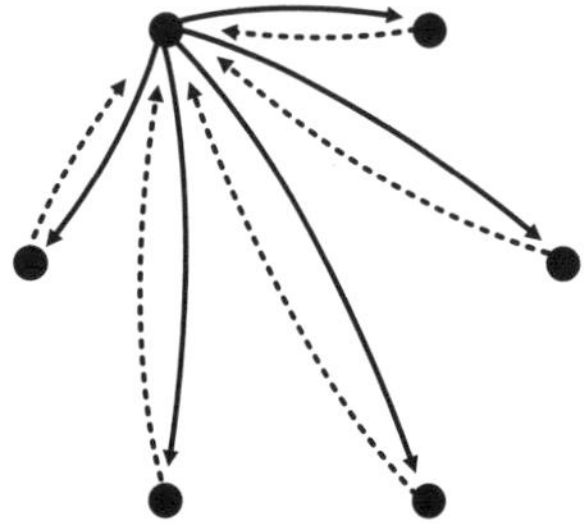

(for `n` Hubsters, where n is at least 3)

Synopsis: `n-note` chords will be chosen by group processes, and smooth or quantized curves of MIDI Note and Pitch Bend data will be computed by me and sent to each player to accomplish coordinated movement between the chosen chords.

To play in this piece, you must have a sound-producing device capable of responding to MIDI Note and Pitch Bend commands, and the sensitivity of its pitch bend must be able to be calibrated to ± 1 semitone.

At varying intervals of time, I will send out a request, in the form of a MIDI `Start`. When that request is seen, each Hubster should promptly choose and send me the following information:

→ a vote for the next tonic (MIDI Note On in the range of `1-127`)

→ an interval from whatever tonic gets chosen; send this as the velocity byte of the tonic Note On message, and pick a number between 1 and 18, representing intervals in increasing dissonance (e.g., `1 = unison, 2 = octave… 18 = 25/24`).

→ a glissando quantization value `(0 = smooth, 1-100 represent divisors of total transition-to-next-chord time)`. Send this as the data byte of a Program Change message (e.g., `C0 33`).

After I have received this information from each player, I will compute Note and Pitch Bend data to send back to each player so that we will all converge on a new chord, the intervals of which have been specified by each of you. The path of travel between each player's current note and his target note in the chords is a function of the total travel period (set by me for now) divided by the quantization. If you send me `0` for quantization, I will compute the smoothest path (in Note Ons and Pitch Bends) from your current pitch to the target pitch in the next chord, and send out those Note Ons and Pitch Bends to you at the appropriate time. Other quantization values will cause more stepwise changes in your pitch—a value of 1 will cause an instantaneous jump to the target pitch; a value of 2 will cause you to spend half of the period at the current pitch, and half at the target; 5 will cause a 5-step change, each step taking 1/20th of the total period, and so on.

You may not play any notes or make any pitch bends directly to your own sound-producing pile of gear. However, program changes or timbre modifications are okay. Tasteful post-processing is acceptable if not done so as to destroy the pitch content of the notes played. Generally, timbre should be chosen so as not to disguise pitch content. Each Hubster should control his volume so as to blend with the whole (ulp!): If you can hear yourself clearly, you're too loud. We will have to make sure we are in tune. I am deeply sorry about this.

Chord freezing: if, at any time, you hear a chord you like, send `MIDI Stop`, and the chord will be held; if, at any time, you wish a chord to be unfrozen, send `MIDI Continue`, and the transition to the next chord will begin immediately. In the absence of any freeze or unfreeze requests, chords will be held for a (relatively short) sustain period chosen by me. One requirement for freezing/unfreezing: these events should only be generated by discretionary action (e.g., a button push, a key press, whatever) NOT as a result of stochastic processes. In other words, freeze a chord because you like it, not because the Tao says it is time for the chord to be frozen.

If our various sound-producing gear does not respond linearly to pitch bend, pretend I didn't write any of the above.

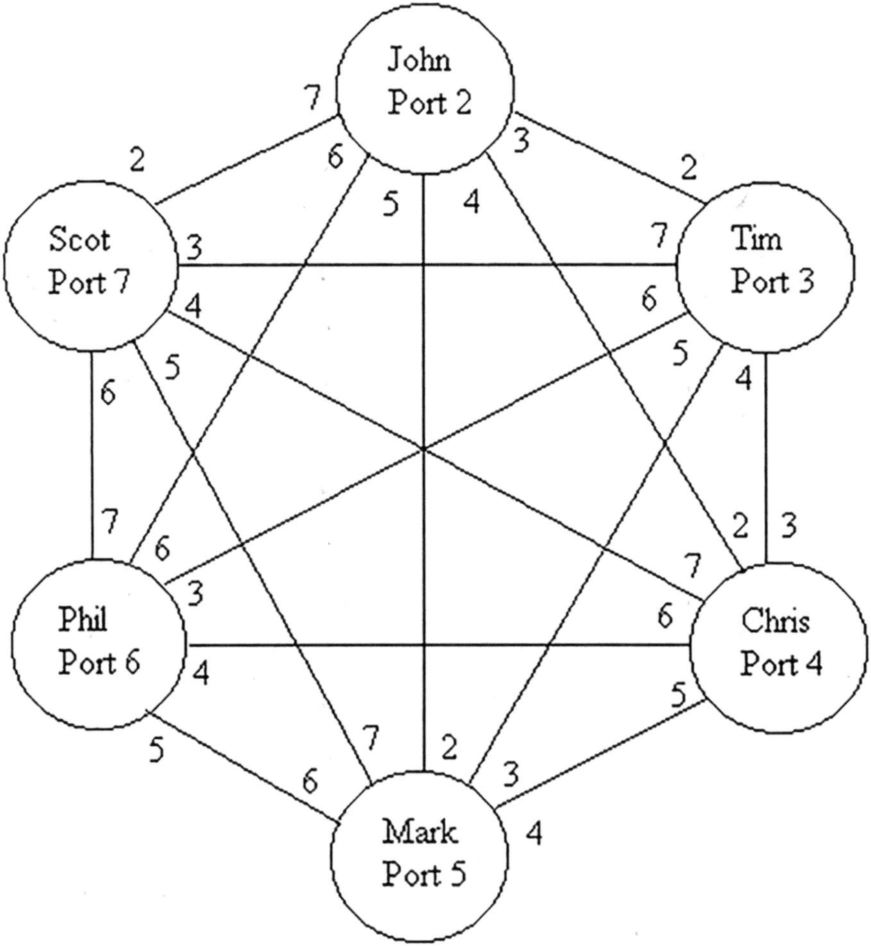
John
Port 2
Tim
Port 3
Chris
Port 4
Mark
Port 5
Phil
Port 6
Scot
Port 7

STUCK NOTE

Scot Gresham-Lancaster 1994

Released on *Boundary Layer*

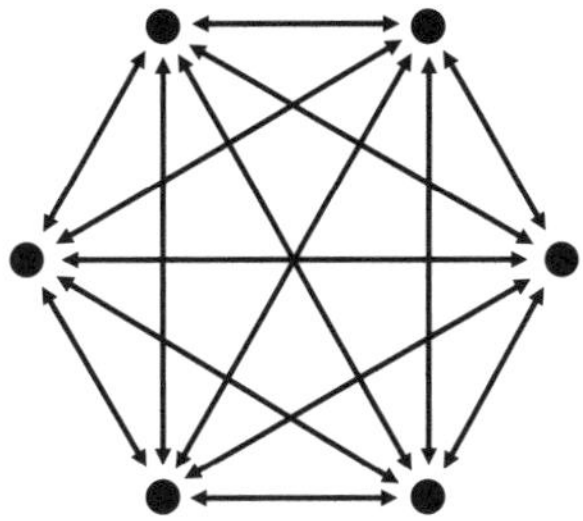

Anyone writing computer code for MIDI synthesizers will have experienced the "stuck-note" phenomenon: a given Note On message is received, but its corresponding Note Off is lost and it becomes a "stuck note." In this piece every player can play only one note at the beginning of the piece, meaning one continuous sound, and that note stays "stuck" on for the duration of the piece. There are only two allowable controls for changing that sound as it plays: a volume control and an "x-factor," which is a controller that in some way changes the timbral character or continuity of the sound. These two controls are always available to be played remotely by any other player in the group, but are never playable by the player who started the note. Players send streams of MIDI controller messages through the hub to other players' synthesizers, taking over their sounds with two control streams. All players are together shaping the whole sound of the group, and an interesting social and sonic situation develops when more than one player contests the same controller, resulting in rapid fluctuations between the values of parameters sent by each. The concept of this piece comes from a conversation with network music pioneer Jim Horton.

One of the last pieces written for the MIDI-based hub hardware. Each player has only one continuous sound, but can vary the loudness and some other factor—completely unspecified—of everyone else's sound.

Stuck Note is the only Hub piece that has been realized by other live computer-music performing groups. The simplicity of the spec lends itself to adaptation and pedagogy.

Gresham-Lancaster adds that "several recordings I have heard over the years drifted away from the 'drone' quality of the original. In a few recent instances, the piece has been played with a 'beat' going which was both disconcerting and sort of weirdly interesting. It showed that the looseness of the score itself leaves it open to both interpretation and subversion. The context of controlling others and not yourself is powerfully different from the normal music-ensemble experience."

VARIATIONS II

John Cage
Hub realization
November 17, 1995

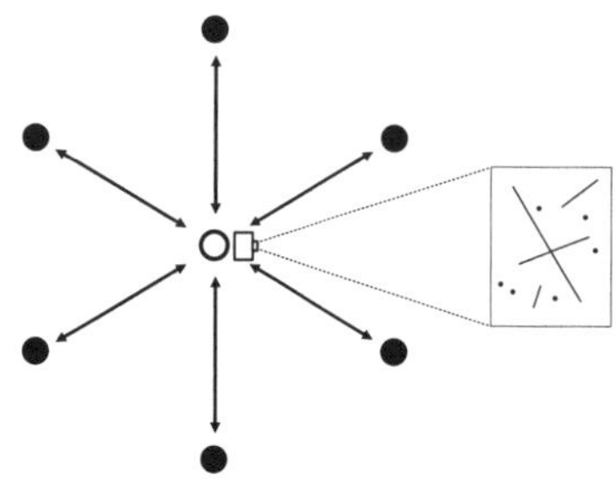

In 1995 David Bernstein produced at Mills College "Here Comes Everybody: A Conference on the Music, Writing, and Art of John Cage." One of the concerts was devoted to realizations of Cage's live electronic music, and the Hub decided to implement a real-time version of *Variations II.*

Variations II (1961) is one of a group of eight works titled *Variations* composed in the two decades between 1958 and 1978. Rather than conventional musical scores, these works are sets of instructions that specify disciplined activities to prepare a performance. The instructions for *Variations II* include transparent sheets that contain either lines or points. The performer(s) are instructed to superimpose these transparencies on a surface and "drop perpendiculars from the points to the lines" to determine readings for six variables of the music. These variables are frequency, amplitude, timbre, duration, occurrence in time, and structure of the sound event. From this (rather tedious) preparation, the performer obtains a score from which the performance is rehearsed.

The graphical algorithm defined by Cage in *Variations II*, with its painstaking method for making decisions about musical parameters, seemed to be ripe for adaption to our algorithmic, network music. We decided to make a "live" version of this work, in which the casting of superimposed lines and points would occur as part

John Cage's original score is described thus:
"The materials for this work consist of five small transparencies with a single point on each, and six larger transparencies with one line on each. These may be superimposed in any way, creating a configuration of points and lines, to be interpreted as a way of producing sounds. This is achieved by dropping perpendiculars from the points to all of the lines (creating a total of thirty perpendiculars). Measurements of the perpendiculars are then used to obtain readings for frequency, amplitude, timbre, duration, points of occurrence within the established period of time, and structure of the event (number of sounds)."[10]

Phil Stone recalls: "Bill Thibault wrote a graphics program/algorithm which did the random 'casts' of lines and dots (specified by Cage), converting the results into data which he fed the Hub, which we then used for triggering/modulating our various systems."[11]

John Bischoff recalls: "As Bill T. generated and displayed score 'castings' (lines and dots), each Hubster automatically received a series of measurements of the resulting perpendiculars. So information went from Bill to us—from score to players—as the main flow. Or so we thought. But we individually could also send a 'nudge' back to Bill asking for more data, i.e., another casting to get more measurement information. This feature was primarily instituted because of the 'structure' parameter in Cage's score, which determined the *number of events*. The higher that number, the more castings one would need to fill out the parameters for all the events. Seeing the video of Bill's animated score after the concert was a shock—casting after casting was being generated, about 4 or 5 a second. So as things turned out, we were all nudging him pretty constantly and the information flow went from the Hub to Bill just as much."[12]

of the performance, the music being automatically computed through the network according to Cage's instructions. A central MIDI interface, the Opcode Studio 5, makes possible the simultaneous distribution to all musicians of measurements made from virtual lines and points, which are video-projected as part of the performance. Each musician computes his own performance independently, but all the data used in these computations arises from the same graphical matrix. In addition, as each musician requires more information for creating new events or for determining the details of a complexly structured event, he may request the graphics computer to "nudge" the overlay, thereby creating the necessary data.

In contrast to the "traditional" compositional method, this realization allows us to create music that is unique to each performance, and that need not be rehearsed; both goals served for us to update Cage's concept to the current sociological and aesthetic situations. The process of realizing the work in new form has re-emphasized to us the appropriateness of his compositional strategies to the medium of electronic music.

One of the strands in the musical philosophy of the Hub was the interest in defining musical processes that generated, rather than absolutely controlled, the details of a musical composition. An acknowledged influence on this interest was the work of John Cage, and it seemed a natural extension to us to try to automate the indeterminate processes used in his work. Many of these processes are extremely time-consuming and tedious; and given that Cage was himself involved for a long time in live electronic performance, we felt a real-time realization of these processes during the progress of a performance was not only feasible, but aesthetically implied.[9]

9 John Bischoff and Chris Brown, Ibid.

10 "Variations II," John Cage Trust, https://johncage.org/pp/John-Cage-Work-Detail.cfm?work_ID=234.

11 Phil Stone, email.

12 John Bischoff, email.

FADE

John Bischoff
April 18, 1997

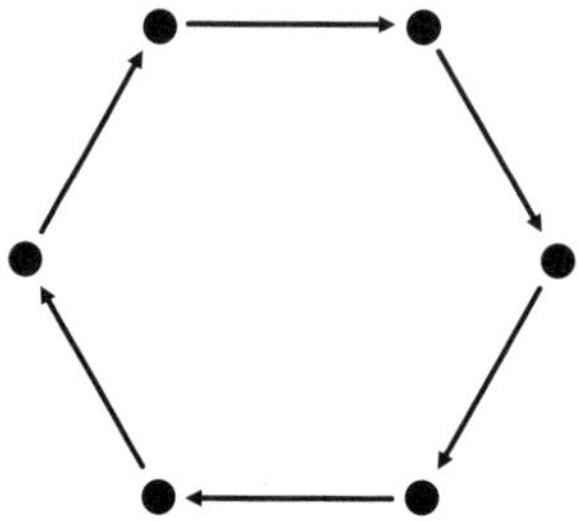

In electronic music performance there is often a tendency to multiply manual controls for a performer, with the thought that more control over more things leads to a more finely crafted music. This piece takes an opposite approach by limiting manual controls to one virtual fader or slider that both drives a player's own system output and shapes attributes of a neighbor's output as well.

For this piece, each player creates a continuous fader-type control that directly plays and shapes the moment-to-moment sound of their instrument. To accomplish this, define at least three lookup tables—for example: one for pitch, one for amplitude, and one for timbre—that translate fader input values (`0-127`) to appropriately scaled values for each parameter. The directionality of correlation is open. Once the lookup table structure is built, define at least six sets of possible functions to fill your tables with so that functions can be switched during performance. Functions can be designed in any fashion, either related to each other or not.

MESSAGING
`fader <integer>` where the variable equals `0-127`.

In performance, your current fader value is passed in "ring" fashion to your neighbor: player 1 to player 2, player 2 to player 3, etc. Each player maps their neighbor's fader values in such a way as to select which set of functions are in use at any given moment.

Bischoff adds: "Given that *one* manual control drives each player's sound—leaving the data exchange out of the picture for the moment—then at a minimum some kind of built-in scheme for pitch and velocity choices would need to be invented by each composer. I envisioned these as mappings of the `0-127` fader value as it is activated in performance. One could add other parametric "maps" as well, of course. Simplest case would be: `fader = 0 = pitch value 0 and velocity value 0`. Each composer is free to use the incoming fader value to influence their mappings in any way they dream up, the idea being that the result would be audible."

LUV CONNECTION

Phil Stone (group collaboration with a CGI / Web implementation by Stone) 1997

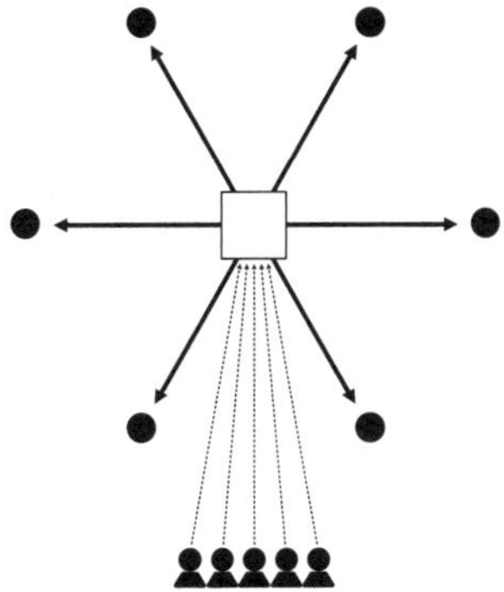

Audience members with laptops access a Web page which gives a choice of 26 buttons, one for each letter in the alphabet. A button press is sent directly to a Hub member distributed in round-robin fashion. Every time a participant pushes a button on the browser, that button triggers a specific sonic event on a Hub member's computer. Hub members have no local control over events; all sounds are initiated by the audience.

A unique Hub piece, controlled by the audience. In its premier at Georgia Tech, in an auditorium that had an Ethernet connection at every seat (in the days before Wi-Fi), it proved to be a failure, as the audience insisted on mashing their buttons as often as possible, resulting in undifferentiated cacophony—an inadvertent proof that any musical system requires sensitive players. The composer comments: "Perhaps it was not so much a failure on the audience's part as it was mine. Working out the website-CGI programming which connected the audience to the Hub took so long, I neglected to design some guardrails into the piece to prevent the 'tragedy of the commons' that ensued." Gresham-Lancaster sees similarities to *Hub Renga* and comments: "My feeling is that with more curation, rehearsal and attention to the 'public' participants it might be possible to get a more musically fruitful outcome."

ETERNAL/ APERTURE

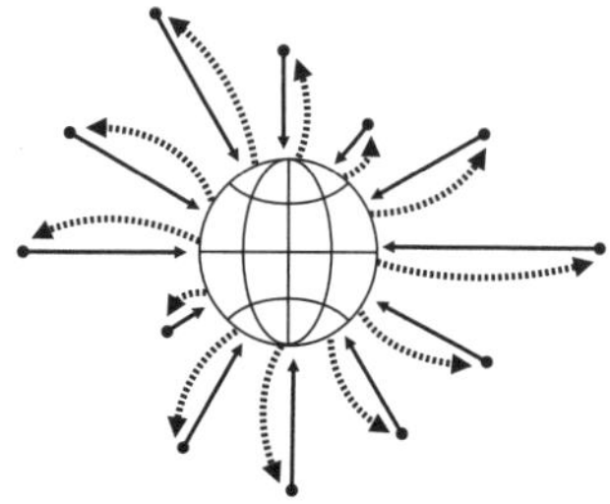

Chris Brown/ John Bischoff 2001

Premiere: July 25, 2005 at Tesla Institute, Berlin

These are two pieces that were performed together by the Hub at Tesla Institute in Berlin in 2005 directly from Phil Burk's TransJam website, written in Jsyn, where they could be played by anyone visiting the site.

In Brown's *Eternal*, up to four players at a time change the harmony and mix of eight modulated sine-wave oscillators, producing a richly vibrating drone. *Aperture*, by John Bischoff, invites sonic exploration and play using a palette of textured noise. By executing mouse-based motions, players can shape large and small changes in the sonic fabric.

These two pieces were part of a concept meant to make network music pieces accessible to the online public, as described below in an unfunded grant proposal from 2001:

"... an ongoing Computer Network Music website which will allow an Internet audience to experience Computer Network Music, both as listeners and as active participants. Computer Network Music is a medium pioneered in the San Francisco Bay Area by such groups as 'The League of Automatic Music Composers' and 'The Hub,' in which automated electronic music instruments interact with each other through computer networks, leading to emergent sonic behaviors that reflect the interdependency of their systems, as well as the interaction of individuals within the group. A repertoire of new pieces is being created for the Internet, each with different sonic and interactive character, and client/participants will select at any time from a growing menu of compositions created by different composers to listen to and interact with.

The music on the website will change continuously like a live sculptural installation. The algorithms continue indefinitely (conceptually, eternally), slowly evolving, and tuned-in by clients via the website. Clients can also play by manipulating graphical controllers on the website, which will audibly influence parameters of the music. The behavior of the music can be influenced at the same time by any participants tuned into the piece, but also by a history of past participation in the piece. The emergent sonic behavior of each piece is thus controlled in part by the present, and in part by the past interaction on the network. The sound reflects the ongoing structure of each algorithmic composition, as well as the level of traffic and interaction on the site. The music is rendered locally at each participant's machine by a DSP instrument which is downloaded from the website to each client machine. Very rich, rapidly changeable, high-fidelity stereo sound is created requiring only very low bandwidth streams of control data, without the necessity of audio data streaming."

HUB 3.0

The band regrouped in 2004 for its third period. The new hub was a simple Ethernet router. The band used it to pass messages in Ross Bencina's OSCgroups, an addition to OSC (Open Sound Control, developed at UC Berkeley's CNMAT). This was a far more flexible protocol, and like the previous change it led to a new repertoire. One significant difference from the MIDI era was that all calculations and quantities were handled directly by the computers and did not have to be translated into MIDI's strict integer-based format. The numbers that defined parameters could be real numbers, or "floating-point," sometimes indicated in the specs as `float` or `f`. Another important change was that laptop computers had become powerful enough for real-time audio synthesis, so the use of outboard synthesizers was redundant.

CUT TO RIBBONS

Chris Brown

June 23, 2005

Premiere: June 1, 2006 in Cuneo, Italy.

Released on *Boundary Layer*

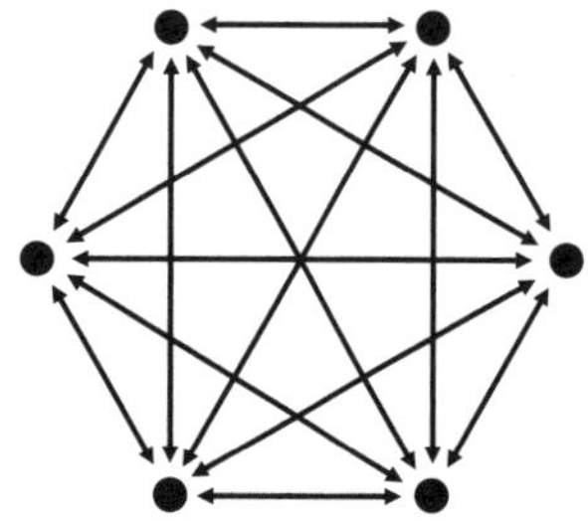

This piece is characterized by shared parameter data targets used by players to stochastically generate phrases, or "ribbons." Each player decides freely when to play each successive phrase. At the start of each new phrase every player sends each target they used in building their phrase to one other player, who saves it for optional use in constructing their next phrase. Most parameter targets are expressed as float between `0.0` and `1.0`; their application to each player's instrument is freely determined. A parameter called "tuning" is broadcast to ALL players at the start of each ribbon, to be used to control the range of stochastic deviation from the targets for their next phrase. The purpose of this system is to create a complex feedback network of data exchange, so that the overlapping ribbons generated are all influenced by each other, but each player retains the freedom to adjust and interpret the data in their own way, with their own sounds, and at times they choose.

INSTRUCTIONS:

Manually trigger phrases of events whose sonic character is chosen stochastically within a range of deviation called "tuning" from target values of the following parameters:

```
pitch   → [0.0-1.0],
          low to high;
amp     → [0.0-1.0],
          silent to loud;
timbre  → [0.0-1.0],
          degree of timbral
          variation;
tempo   → [0.25-40.0],
          events per second;
rhythm  → [0.0-1.0],
          offset from 0 to
          tempo period
          (0.0 = happens
          exactly on pulse,
          1.0 = happens
          anytime within
          the next tempo
          period);
dur     → [0.0-1.0],
          on-time duration,
          proportion of
          tempo period;
density → [0.0-1.0], mute if
          rand(1.0)
          > density;
phrase  → [1.0-10.0],
          duration of phrase
          in seconds.
```

Provide your own GUI controls for choosing targets for each parameter and for selecting one player whom you will send it to after triggering your phrase. Your message should look like this: `/playerName myName param value`. When you receive such a parameter message, you automatically set your GUI control to the value received; you can always change targets manually before using them in your next phrase.

Implement stochastic deviation of these parameters in computing your phrase using a "tuning" variable in the range `[0.0-1.0]`. The "tuning" value sets the range of random choice around the target for ALL parameters. Send your "tuning" value to `/hub` with each phrase trigger: `/hub myName tuning value`. Choose and change manually the player that you allow to set you own "tuning" value. You can always also choose to set your "tuning" value manually.

The composer comments: "As my coding practice became more developed, my ideas for how to structure pieces for the Hub became more complex. I hoped that the challenge of more complexity could achieve a greater variety in the character of new pieces. But since each of us writes our own code, the complexity of debugging the interaction of six versions of a complex spec like this one became prohibitive. Nevertheless, this piece combined multiple layers of network interactivity defined within a context where rules could still be freely followed or ignored."

BOSS

Phil Stone
2005

Premiere: July 5, 2006 at Conservatorio di Musica, Padua.

Released on *Boundary Layer*

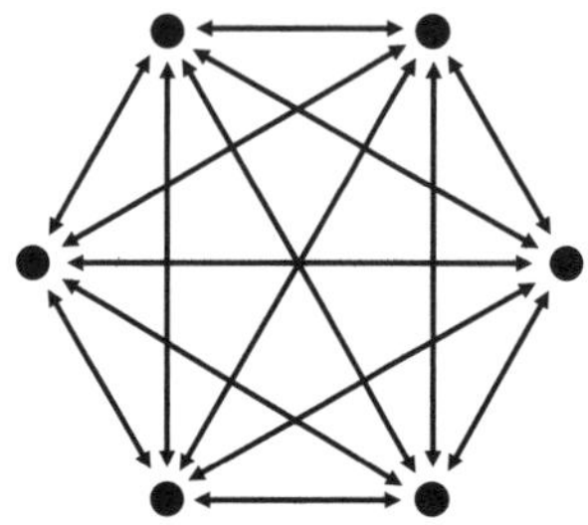

This is one of the first Hub pieces written for the third-generation hub device, a simple off-the-shelf Ethernet router passing OSC (Open Sound Control) messages. From this point on, the band will use the OSC messaging format for passing data of any description and numeric value rather than the slower and more complicated translation of data into and out of MIDI's note-based and strict 0–127 integer format. (The OSC message format is recognizable as a forward slash followed by directives and data, e.g., `/hub parameter value`.)

"PD" is PureData, the music programming language used by Stone. "Royalties to the Cage estate" is a joking reference to Cage's silent piece *4:33*. No explicit mechanism is suggested for informing players of who the current boss is or what the current amplitudes are, though all players would have access to all messages.

The composer comments: "*Boss* might well have been titled *Minister of Amplitude*, considering the debt it owes to Tim's [*Minister of Pitch*] ground-breaking re-apportioning of roles. But it injected a little anarchy into the mix."[13]

In this initial version of the piece, there is one Boss at any given time. Anybody can become Boss at any time, deposing the former Boss.

The Boss will be completely in charge of the Hub mix. The Boss sends `amplitude` messages to one or many Hubsters—including to themselves—in any way desired.

Sending a `trigger` message (value unimportant) to `/hub` effects a coup d'amplitude. Upon receipt of such a message, the former boss must stop sending `/amp` messages immediately. The new boss cranks out `amplitude` messages until similarly deposed.

All Hubsters should be making sound (of any conceivable kind) pretty much constantly; this sound is to be instantly attenuated directly proportional to incoming `amplitude` messages.

A non-Boss Hubster may not make any adjustment to their own amplitude, except to adjust the internal balance of their sound [1], or to achieve Group Balance [2], or in Emergencies [3]. What the Boss chooses to do with this dynamic hegemony is completely unspecified. For example, I hope to try some PD-controlled time-slicing and/or crossfades (something like extremely primitive granular synthesis with the Hub as source). A simple way to get up to spec would be knobs, one per Hubster. The completely minimal version would be to set everybody to `0.0` for the duration of your rule, but that might call for royalties to the Cage estate. It would be wonderful if everyone had different ideas about this.

[1] Adjusting the mix of individual sounds that make up your overall sound is perfectly ok, just make sure the final output amplitude is controlled by the Boss.

[2] Everybody should make sure their setup, sent an amplitude of `1.0`, is not going to explode any eardrums. Resist the temptation to micromanage or actively thwart the Boss's will, though.

[3] Gear glitch, death in the audience, Homeland Security Condition Mauve or higher.

13 Phil Stone, "Non-Mathematical Musings on Information Theory and Networked Musical Practice," *Organised Sound*, in press.

TESLA SYNC

John Bischoff
2005

Premiere: July 25, 2005 at Tesla Institute, Berlin.

Released on *Boundary Layer*

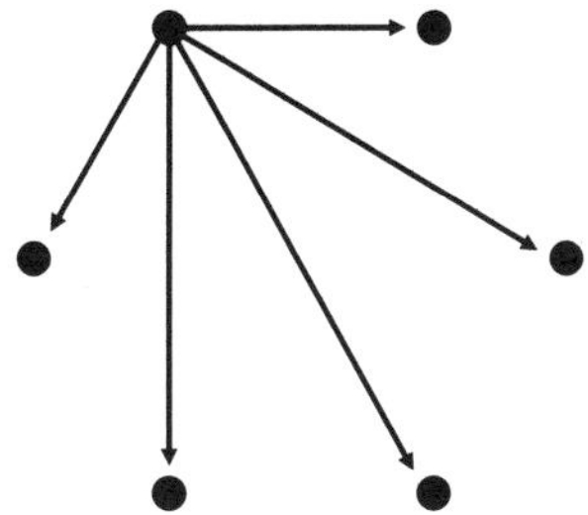

Compare *Wheelies*, which also provides a single stream of triggers to all players. In this case the "continuous stream" of triggers was a quasi-regular pulse that varied between one per second and ten per second, with a general trajectory of increase.

One player establishes a rhythmic framework by generating a continuous stream `/trigger` messages via OSC to all players, including themselves. The framework represents all possible event onsets for the piece. Players respond to some percentage of triggers and ignore the others according to their own adjustable variable played in the moment. As players vary their playing percentage independently, complex and unpredictable rhythmic patterns are created which always remain synchronized at the lowest level of rhythmic action. Player sound qualities can be changed at will as the piece unfolds. The player who has the role of trigger generator can vary the tempo at any time and start or stop triggers as desired.

"Free intonation" is what Lou Harrison called "free style just intonation," a kind of tuning he used in his 1955 *Simfony.* It was Harrison's approach to the unsolvable problem of modulation in a fixed gamut of justly tuned pitches. Harrison simply discarded the fixed gamut, and let every new pitch be the starting point for further extensions of an unlimited gamut. One radical consequence of this approach is that the "tonic" is no longer fixed, but drifts.

Perkis: "While Harrison's invention of 'free intonation' could never become practical with material acoustic instruments, *Lou Drift* shows that such a practice is trivially simple using computer-based instruments."

LOU DRIFT

Tim Perkis
2005

Premiere: July 25, 2005 at Tesla Institute, Berlin

Released on *Boundary Layer*

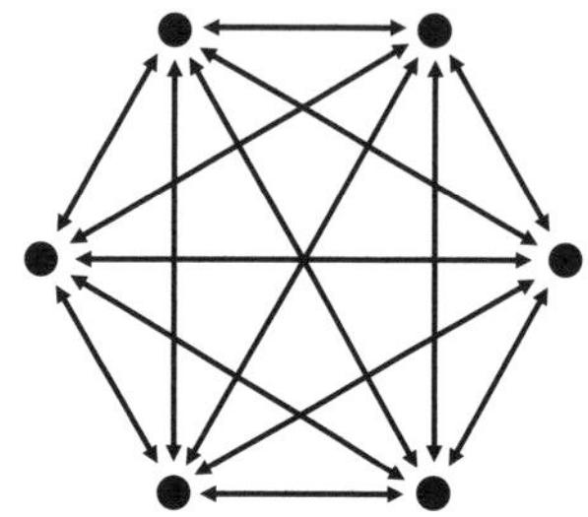

This is a piece in free intonation, in which there is no gamut or scale defined. Each pitch played is calculated based on a simple just rational relationship with another currently sounding voice.

The piece is very much like *Waxlips* in slow motion, but simpler and more restricted.

The basic behavior requested: pay attention to a `/pitch` message which is sent to you (you will not be paying attention to EVERY `pitch` message you see: more on that later). Interpret the value of the `pitch` message as a real number frequency in Hz. You should play and hold a reasonably long tone (several seconds long at least) at the frequency requested, and calculate a new `/pitch` message that you send immediately to `/hub`. Calculate the new pitch by multiplying the pitch that you played by one of the set `{2,3,4,5,7}` and then dividing by one of `{2,3,4,5,7}`.

In addition, I will be sending occasional `/density` messages. Take the number provided as a minimum time, in seconds, that you pause after you perform a `play/calculate/send` action before doing it again. After your pause deadtime is over, grab the next (or a very recent) `/pitch` message and use that as the basis of your new action.

You need to be able to catch any sent `/density` message sent, even if it is sent during your pause time.

PINS & SPLITS

Mark Trayle
2005

Premiere: July 25, 2005 at Tesla Institute, Berlin

Released on *Boundary Layer*

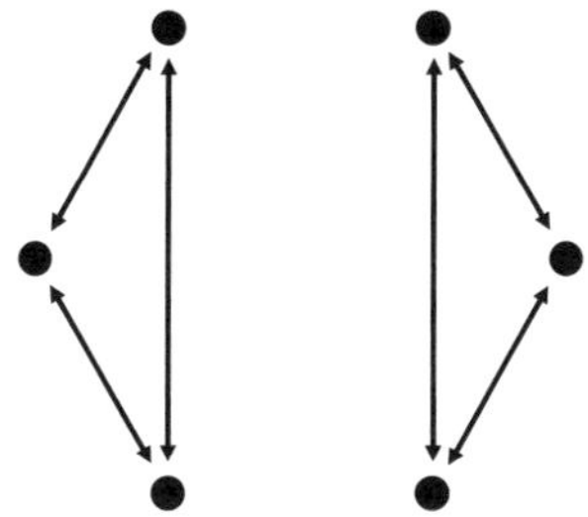

In a program note, Trayle wrote: "The musicians alternate between a relatively still background sound and a foreground improvisation. What they play is up to them, but when they play is determined by another player. The ensemble will fluctuate between mischievous disruption and peaceful cooperation. And no, the title is not a reference to bowling. The title morphed from the original in an email thread with my fellow Hub-mates."

0. divide the ensemble into two groups.

1. make a sound. when you receive a `/trigger+/density` message start a new sound and apply the `/density` value to it in any way you see fit. this sound should stop after a while and you should go back to the first sound you made.

1a. you can send `/trigger+/density` messages to anyone in your group at any time you wish. you should have direct control of `/trigger+/density` messages, i.e., a performer interface rather than an algorithm.

1b. you can react to `/trigger+/density` messages with as many different sounds, sound-events, sound-aggregates, etc., as you like, but always return to the first sound.

2. variation 1: add a `/dur` parameter to the message and play sounds only as long as indicated in the `/dur` value.

NOOSPHERE: Sonification of Global Consciousness

Scot Gresham-Lancaster 2005

Premiere: July 25, 2005 at Tesla Institute, Berlin.

Released on *Boundary Layer*

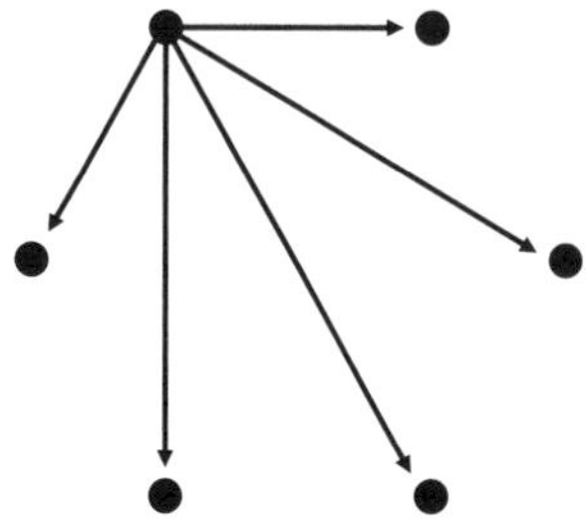

I am spewing `/yourname scot timbre n` **where n is** `0.0 > n > 1.0`
do with it what you will

The tersest Hub spec.

Gresham-Lancaster explains: "The Global Consciousness Project (http://noosphere.princeton.edu) has been collecting data from a global network of random event generators since August, 1998. The network has grown to about 65 host sites around the world, each running software that reads the output of physical random number generators and records a 200-bit trial sum once every second, which is sent to the Princeton site. This piece is a sonification of the state of the network at the time of performance. Each performer is sent one piece of data in sequential order from each of the sites around the world that are currently part of the grid. This is a floating-point number between 0.0 and 1.0 to be used to control timbre. The piece is similar to *Stuck Note* with this parameter representing what was called the 'x-factor' in that piece."

SWAP POINT

John Bischoff

July 1, 2006

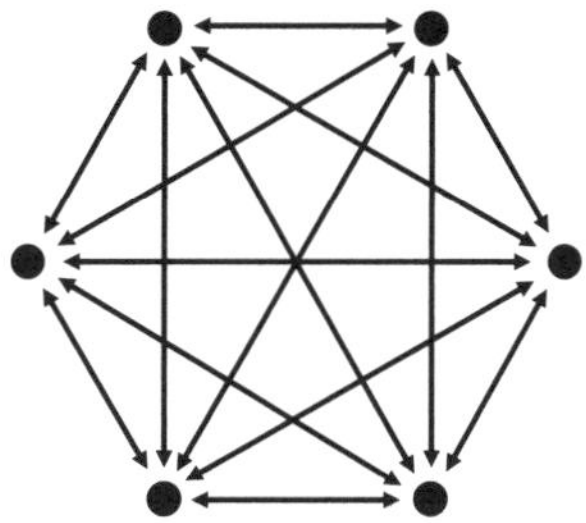

This piece features acoustic sound recordings expressed through smoothly moving grain players (as in "granular synthesis") which are diverted away from their smoothness by incoming Hub data.

Specification for the smooth grain player (ranges indicated are ballpark suggestions):

→ record 10-second-long sounds of acoustic or electroacoustic origin and save them as sound files
→ build a sound-grain player to play grains of your sound files in sequential fashion following these guidelines:
[1] every time a grain is played move the location of the grain stepwise forward or backward 1 millisecond within the sound file
[2] make the play rate of the grain player variable, 100–0.25 Hz
[3] make the grain duration variable, 50–200 milliseconds
[4] fix the grain pitch at the original pitch of the sound recording

DATA MESSAGES:
→ `/trigger <float>` (0.–1.0) = alternate grain location within the sound file
→ `/pitch <float>` (.125, .25, .5, 2., 4., 8., 16.) = octave multiplier for alternate pitch of grain

DATA IMPLEMENTATION
→ when a `/trigger` message arrives, defeat internal grain-player for 10–100 milliseconds (variable) and use the trigger value to play a grain at the specified alternate location and at last specified alternate pitch according to most recent pitch data.

PERFORMING ACTIONS
→ start/stop grain player
→ vary grain rate
→ vary grain width
→ vary direction of grain movement within sound file
→ vary defeat duration
→ switch current sound file
→ send `/trigger` messages and `/pitch` messages to other players; triggers can be sent at any rate and under the guidance of any manual or automatic mechanism; pitches should be sent at a rate `=< triggers`.

Granular synthesis is a technique that plays back a prerecorded sound sample as a series of "grains," or short segments, each usually lasting a fraction of a second (Bischoff requests 1/10 to 1/100 second grains). By varying the grain width, the rate of playback, the direction of movement, the degree of overlap, and other parameters, the original sound can be played back unaltered or greatly modified—stretched or compressed in duration, raised or lowered in pitch, played backwards, etc.

A site-specific composition for an event at Deutzer Brücke in Köln, Germany. The entire concert took place inside a hugely reverberant tunnel supporting this 370-meter bridge over the Rhine.

TACOMA OSCILLATIONS

Mark Trayle 2006

Premiere: July 11, 2006 at Deutzer Brücke, Köln

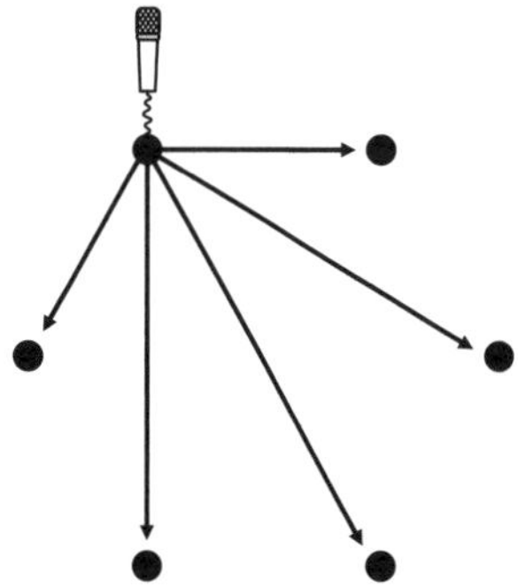

i came up with a max patch to take the microphone feeds from the bridge, do a bit of analysis on them, then distribute the results. this isn't really a spec for a piece, it's a way to get data to use in whatever patch/code you'll be running for our cologne gig. i'm imagining that we'll play some selections from our repertoire, and then do a big improv with these bridge signals.

there are six mics on the bridge. each of these signals is analyzed for noisiness and its amplitude envelope is followed.

you get
envelope follower data `(0. to 1.)` from `/hub` as `/amp`.
you get "noisiness" data `(0. to 1.)` from `/hub` as `/timbre`.

you can
choose which set of signals to listen to by sending `/hub` a `/trigger` message where the values are `0.0` = channel 1, `0.2` = channel 2, `0.4` = channel 3, `0.6` = channel 4, `0.8` = channel 5, `1.0` = channel 6. so if you send `/hub you trigger 1.0` you'll be receiving envelope and noisiness from mic #6.

you can
change the smoothing of the envelope follower signal by sending `/hub` a `/tuning` message in the `0` to `1` range. `0` is very bumpy, `1` is pretty much a flatline.

you can
change the smoothing of the noisiness signal by sending `/hub` a `/density` message in the `0` to `1` range. `0` is very bumpy, `1` is pretty much a flatline.

using some sound files that Hans posted i've found that envelope changes often, whereas noisiness is more stable. sounds on the bridge are pretty noisy, as you'd imagine. maybe one of the drunken soccer louts will find a mic and sing their national anthem, which would reduce noisiness considerably.

HOT POTATO

Scot Gresham-Lancaster
2006

Premiere: July 11, 2006 at Deutzer Brücke, Köln

Released on *Boundary Layer*

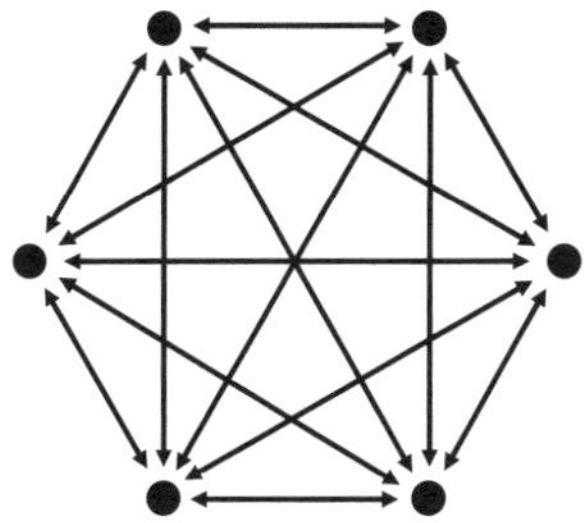

I have worked out the spec of a new piece if anyone is interested in such a thing. In keeping with my tradition of ultra-simple pieces this one is called *Hot Potato*, like *Waxlips* but you have to play a note (any note or sound) before you randomly throw the single potato. The potato is simply a number that sets the duration of the sound event in each instance (i.e., `100 = 100ms, 200 = 200ms,` etc.). Any potato handler can change the number… or not.

This piece is like *Waxlips* in having a one-in-one-out structure; that is, a player receives a message, makes a sound, and passes a message to another player. However, the network is much simpler, as duration is the only parameter passed and there are no defined rules or limits on its passing—a musical situation that opens a tremendous range of possibilities, requiring great sensitivity for a good performance.

Gresham-Lancaster comments: "This was a rollicking good time to play when we finally got it all running. One time in Cuneo, Italy we did a version with 18 students playing with us in this beautiful old church. We spent a week workshopping it and getting everyone to find their voice. It was that performance that eventually was released on the CD. The experience of that workshop gave me an insight into the difference between the 'laptop orchestra' model of homogeneity and its vestiges of the classic orchestra hierarchy and the League/Hub anarchic heterophony that is so distinctive and has not fully survived the transition to an academic context."

V'AV

Chris Brown

January 1, 2008

Premiere: October 23, 2008 at the Contemporary Jewish Museum, San Francisco

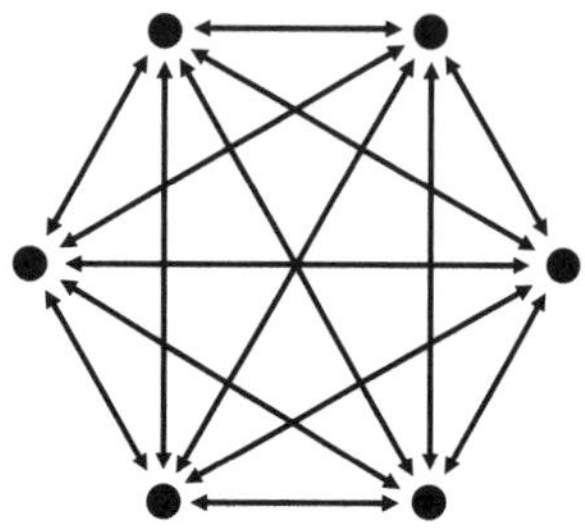

GENERAL DESCRIPTION

Players generate chains of sound events by playing a sound and, after a variable delay time, sending another player the value of one parameter used in their synthesis algorithm, which triggers the receiver to play a sound also using the received value. Shared parameters are thus used to produce sounds that evolve through group interaction. Any player can stop or start a new chain. Each player sends their mono audio signal to one player (the conductor) who moves them in multiphased 3D trajectories using an Ambisonic mixer. This slowly evolving texture is occasionally contrasted by players becoming "guerrillas" who improvise freely, but whose virtual spatial position is fixed. The conductor can also produce coordinated changes by seeding the group with a common set of synthesis parameter values, and globally setting shared variables of fundamental pitch and delay tempo.

SPECIFICATION

Each player designs two synth instruments: one percussive, and one continuous—their timbres should be able to change from a pitched to a noisy quality, and include varying amounts of reverb, chorusing, or other thickening effect.

Each player must be able to automatically send and respond to `control` messages which are sent to and received from one other player. These messages transmit values used by the sender of one of the following six parameters:

`pitch` → integer from 1 to 32 that sets synth frequency to one of the first 32 harmonics of the global variable "fundamental" (set by conductor, ranging from 20 to 320 Hz);

`amp` → `[0.0-1.0]`, sets amplitude;

`instr` → `[0.0-1.0]`, sets probability of a percussive or continuous instrument;

`noise` → `[0.0-1.0]`, sets an instrument's timbral character from pitch to noise;

`wet` → `[0.0-1.0]`, sets a dry → wet reverberation effect;

`delay` → integer from 1 to 100 that multiplies the global variable `tempo` (set by conductor from 1 to 60 per minute) to set wait time before sending a new `control` message to another player.

The conductor occasionally shuffles and sends new assignments of each parameter to one other player as part of a `cue` message. Each player then controls the evolution of their parameter by transmitting only its value in `control` messages it sends to other players. Each player develops their own algorithm for changing the value of their parameter, which must always be changing, and unconstrained random choice is discouraged. The `cue` message, sent only by the conductor, also fulfills two other functions: 1) it sets everyone's control parameters to the same starting values; and 2) it sets the values of the global `fundamental` and `tempo` variables used by players in interpreting `pitch` and `delay` messages.

Upon receiving a `control` message, the receiver must immediately play a sound that implements the last values of the five parameters they received, as well as the current value of the parameter they control. There is no manual aspect to this playing process—it should be completely automated, using the latest values of control parameters to make a new sound. But once a sound is produced, it can be tweaked manually, and a manual volume control is advised. After waiting the delay-based wait time, the player sends a new control message with a new value of the parameter they control to another player. But players can choose to stop an event stream manually which disables a new `control` message in response to the next `control` message they receive.

At any time a player can request guerrilla status by sending a `guerrilla` message to the conductor. If granted, this means they stop playing in response to received `parameter` messages and instead improvise freely. Guerrillas must however continue to generate `control` messages in response to receiving them. Guerrillas play until they resign using a `guerrilla` message, or upon receiving any `cue` message, when all guerrillas must revert immediately to normal player status.

The conductor generates a trajectory envelope for 3D spatialization of each player's sounds, but players can change their speed of motion through this trajectory by sending the conductor a `speed` message with a `0.0` → `1.0` float meaning slow to fast.

OSC MESSAGES

cue: `/cue your-control-name fundamental tempo parameters`

control: `/control parameter-name`

guerrilla: `/guerrilla` `<<in` **or** `out>` **or** `<yes` **or** `no>>`

speed: `/speed`

INTERFACE

Display the global `fundamental` and `tempo` variables.

Display last values received of six control `parameter` messages.

Display current parameter assigned to you.

Interface to adjust the algorithmic generation of control parameters, and display the last value you sent. (One set of controls is sufficient but must generate pitch and delay values in the ranges given).

`Button` to start a new event stream.

`Button` to stop an event stream. Display guerrilla status (normally off).

`Buttons` to request and to resign guerrilla status.

`Controls` for your own guerrilla instrument.

Commissioned as a sound installation for its opening by the Contemporary Jewish Museum in San Francisco, curated by John Zorn. It is the only Hub piece that was not designed for live performance, although it was itself a recording of a studio performance, edited from one hour to 36 minutes, to be played continuously in a loop.

Conjuring sounds out of the acoustics of a room, this piece owes more to Alvin Lucier's *I Am Sitting in A Room* than to The Beach Boys' *In My Room.* In addition to Trayle's `knock` messages, other players were free to send their own "knocks" whenever they chose. Gresham-Lancaster observes: "Like Mark's first piece *Simple Degradation*, this piece had an understated elegance that pulled the listener in and made its own internal process apparent to any astute listener."

IN MY ROOM

Mark Trayle
2008

Premiere: November 5, 2010 at Queen's University, Belfast

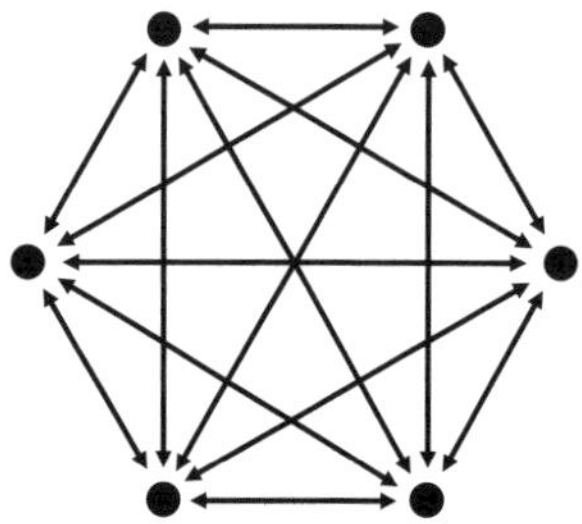

I'm no Brian Wilson, but I'm probably the most tanned member of the Hub. Here's a spec…

Make an impulse response recording of your "studio," whatever room of the house you make your music in. Send me the recording and I'll analyze it, or you can do it yourself. From that analysis I'll pick a set of 5–10 resonant frequencies for you to use in this piece. Everyone will have the same number of frequencies.

Make an instrument that uses those resonant frequencies. As always, timbres are your choice, but they should be timbres that allow those frequencies to be readily heard. Freely improvise with that material.

Be prepared to send and receive a `/knock` message. That message will be followed by an array of the form:

`id → number 1-6` for each hubster duration, explained below

`dur → duration`, explained below

```
f1 → frequency
f2 → frequency
f3 → frequency
f4 → frequency
f5 → frequency
f6 → frequency
```

When you receive a `/knock` message add a voice to your playing that has a different timbre and uses the frequencies in the array passed. (I will figure out how many frequencies will be passed in the array, depending on the analysis of our IRs).

→ Use the "dur" part of the array as the duration to play the additional frequencies. Please make the durations you pass ≤ 1 minute.

→ You should have one distinct timbre for each of the other players. That timbre could be noisier than the one you use for your own resonant frequencies.

→ You should be prepared to play all of your room timbres plus up to five more, simultaneously.

As is our practice, the piece is over whenever we feel like it… probably about 7 minutes.

If you don't know how to make an IR, try this: record a balloon popping, make sure it doesn't clip.

KNOW-NOTHING MUSIC

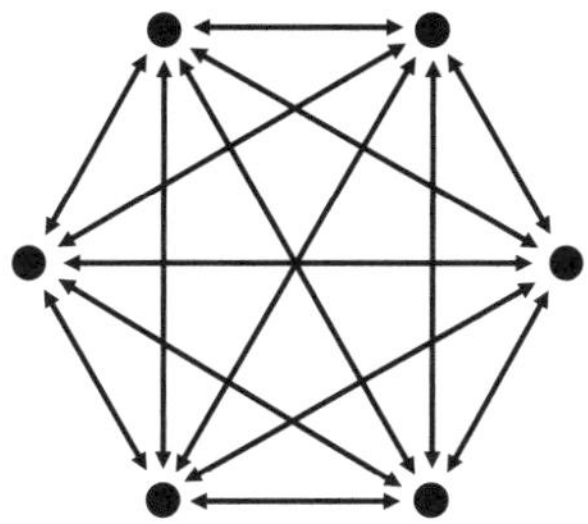

Chris Brown
October 5, 2010

Premiere: November 5, 2010 at Queen's University, Belfast

Free improvisation, but with an instrument you build that automatically tracks one parameter from your music, converts it to a scale of `0.0 → 1.0`, and sends it as the message `/hub me ctl val` where `me` is your name, `ctl` is the name of the parameter, and `val` is its scaled value. Send at whatever rate you like, as long as the parameter is changing. Also send a `/hub me start` message when you start a phrase, keep track of its length, and send a `/hub me end dur` message, where `dur` = length in seconds of the last phrase, when you end it.

Your instrument must have at least five (or one per player) control parameters that affect its sound. Freely assign input control streams from other players to your sound's parameters and re-patch them frequently during the piece. Your instrument should also have the ability to trigger five times processes that change any aspect of its sound. Trigger these processes with any incoming `start` messages, and use durations from any past `dur` messages to set the duration of the process. Initialize all parameter and durational controls to random values to start the piece.

Brown comments: "Not knowing what you are doing is a normally a 'no-no' in writing music, but acceptance of one's own ignorance can actually be quite valuable. A 'Know Nothing' network uses rich specifications, loosely applied. It presumes no necessary behavior in response to data made available within the network, but assumes that performers will have a vested interest in using it. I'm interested in precisely defining systems that create outcomes that are not known in advance to me, even if I made them. Not knowing what happens when we feed stuff into the network is the most important aspect of our practice."

The real-time spectral analysis of acoustic sound upon which this piece depends would have been impossible with twentieth-century laptop computers. The piece continues Bischoff's longstanding "focus on the material nature of the medium." The electronic sounds all derive from the acoustics of physical objects.

PARTIAL GUIDANCE

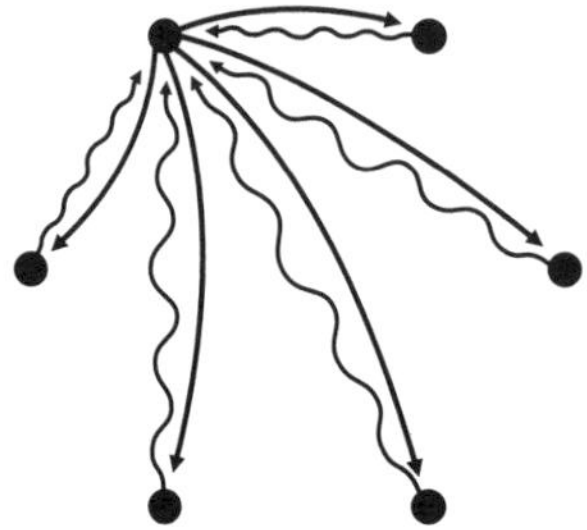

John Bischoff

July 15, 2010

Premiere: November 5, 2010 at Queen's University, Belfast

In performance, players create short acoustic sounds by manipulating small amplified objects. As sounds are made, one computer performs spectral analysis of the acoustic audio and distributes the most prominent frequencies among the players via the hub. Each player's computer responds in some musical way to incoming frequencies derived from the sounds of the objects.

Each object is amplified by a piezo disc that is fed into a separate input channel of the analyzing computer's audio interface. Each analysis frame detects six of the most prominent partials in each acoustic event and randomly distributes one of the partial frequencies to each player, including themselves. Each player will therefore receive a single floating-point frequency number on each acoustic sound made. The `frequency` message will have an OSC label of `/obj#` where `#` is an arbitrary integer representing the source object `(1-n)`.

Frequency messages are sent within a very short time interval after the onset of each acoustic sound (ca. 50 ms) so an immediate response to an incoming frequency will result in near synchronicity between acoustic event and electronic response. The options of a delayed response, or no response at all, are equally valid. Each player will need to find at least one small acoustic sound object to play and build or purchase a piezo contact mic to amplify it.

LIP BOMB

Scot Gresham-Lancaster 2010

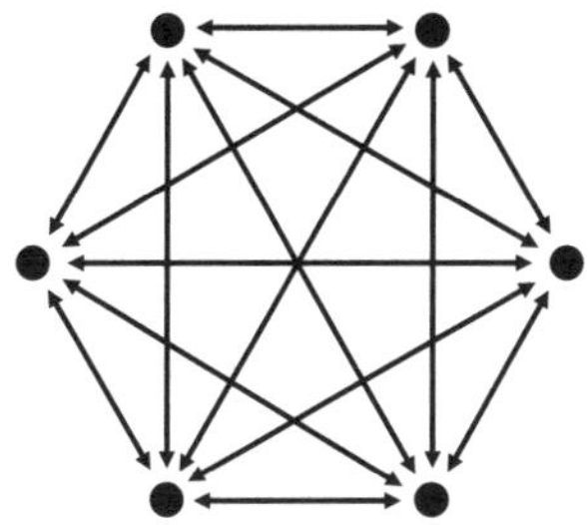

Premiere: November 5, 2010 at Queen's University, Belfast

(6 minutes)

Pick sounds that are no more than 500 ms, predominantly vocal vowel and fixed frequency samples or digitally generated phonemes… you choose… all from the word "terrorist."

pick a single ratio (for example `8/5`, or the ratio is `r = 1.6`)

so the message looks like

```
/scot tim freq 440
```

then

```
1st time out-freq =
in-freq * r
2nd time out-freq =
in-freq / r
3rd time out-freq =
in-freq * r
```

etc., so then

`/mark scot freq 660` (assuming the first time and `3/2` as the ratio)

the second time it would be something like

```
/chris scot freq 293.334
```

(assuming the ratio is `9/4`)

I am "hosing" this *Waxlips* variation with motion detected from a camera pointed at the audience if I can arrange it…

Every minute I am planning to print the state of the Hub from the "processing" state animation on an inkjet printer. I am injecting various human pheromones into the ink cartridges. I will dry each print out in front of a fan that is pointed at the audience.

"Hosing" means seeding the process with starting values, as was done in *Waxlips* and *Vex*. Incidentally, this is another free-style just intonation piece, as the frequencies change by integer ratios with no controlling tonic. As with *Lou Drift*, the newer technology makes this far easier to accomplish than it was for *Listener's Digest*.

Gresham-Lancaster comments: "I had the original inspiration for this piece in relationship to a very personal experience. My youngest son was afflicted with schizophrenia late in his 16th year (18 years ago). He still lives with us and it has been a real challenge. It is a misunderstood affliction that is often not treated as other illnesses because of the nature of the affliction itself. My wife and I have spent all those years researching treatments and trying to understand the best way to manage this very perplexing malady. In my ongoing research I came across a paper from 1960 that identified a distinct if very subtle odor that has been emitted by those with schizophrenia.[14]

I was going to try to synthesize this odor and have the pheromones printed via ink jets to fans pointed at the audience. The idea was to trigger the subconscious reaction to the presence of a psychotic individual as part of the anxiety of this piece about the word 'terrorism.' Not everyone could do the phoneme ask, so we modified it in rehearsal."

14 Kathleen Smith, M.D. and Jacob O. Sines, Ph.D, "Demonstration of a Peculiar Odor in the Sweat of Schizophrenic Patients," *Journal of the American Medical Society Psychiatry* (February 1960), https://jamanetwork.com/journals/jamapsychiatry/articleabstract/487740.

Bischoff's performance setup for this piece includes a counter for each player, a useful addition to keep track of where the other players are in the process.

ONCE

Tim Perkis
2010

Premiere: October 27, 2014 at Mills College, Oakland, CA

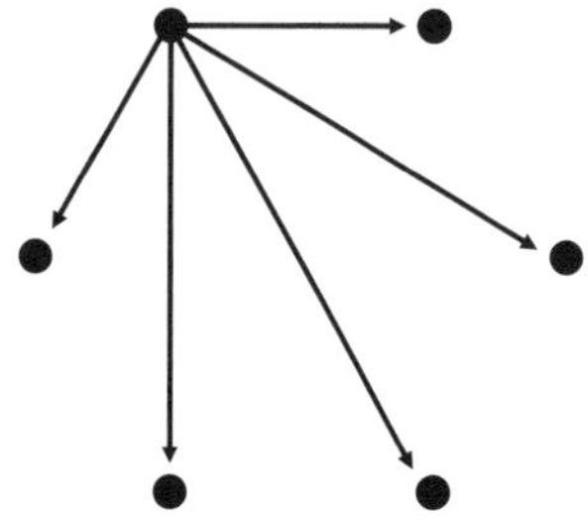

Create a set of 26 sound events, each of which is essentially identical each time it is played, and whose duration is less than 10 seconds. Some of them should be much shorter than that, percussive in nature.

Make an "instrument" for yourself to play these events in which each of these events is assigned to and triggered by an alphabetic key `(a-z)` on your computer keyboard.

Build in a mechanism which ensures that each of these sounds can only be triggered once. Please really build a mechanism which will, within any one run of the piece, disable the key trigger after it is hit once. (In other words, make it impossible to cheat.)

Be able to respond to these OSC `messages:`

`/reset <john, mark, etc>` → when received this should cause all your keys to be re-enabled.

`/play <john, mark, etc> <a-z>` → should cause the sound associated with the given key to sound, even if you've already played it once and it is currently disabled for you to play from the keyboard.

Playing action: freely improvise during the piece. In the simplest form, the piece ends when we've all used up all our sounds.

I'm imagining it as another of these spacious Feldman-esque deals. You might imagine some particular role you want to play in the emerging sound landscape and give your sounds some consistency based on that idea: a tuning, a timbral family, whatever comes to mind. Or not; this is completely optional.

I'm not sure if I'll ever send any `/play` and `/reset` messages. If the manual one-pass, free-improvised version ends up sucking I might be sending automatically generated `/play` messages and or `/reset` messages. The `/reset` message will more likely just be available as a convenience to allow the piece to be started and restarted from a central location (me).

Once (TP)
J. Bischoff - Sept. 2010
6:00
minutes!
once master
reset player with \ (backslash)
set player w/ 1-5
1 john
26
2 mark
24
3 phil
0
4 chris
26
5 scot
24

FENCES

Matt Ingalls

2014

Premiere: October 27, 2014 at Mills College, Oakland, CA

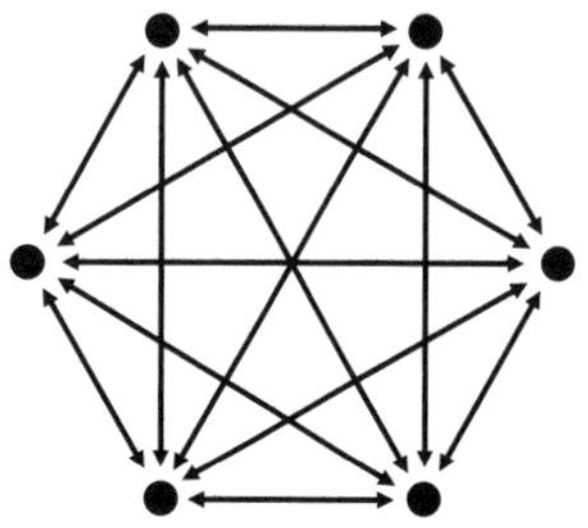

A "port" of a few passages from acoustic pieces I have written for sfSound—it's basically a Wolff-inspired, "hot potato" piece. There should be lots of silence and quasi-unison chords so it might be kind of dramatic for the audience—and hopefully enough "manual" playing to make it fun for us. (In this spec, I use "manual" to mean something *you* do during the performance and "automatic" as something your software does during the performance.)

SOUNDS

You need to produce two kinds of sounds:

`+ Posts`

Short and percussive sounds that should be pitched when responding to `note` messages, but may be pitched or unpitched when responding to `response` and `stop` messages. Generally loud, but dynamic variations are OK. Will need to be triggered manually and automatically.

`+ Slats`

Long sounds that can be pitched and/or unpitched, used when responding to `response` messages. They need to be able to sustain indefinitely, until signaled to be turned off. Generally soft, but dynamic variations are OK. Will always be triggered manually (but will be turned off automatically). At the beginning of the piece, the sustained sounds should be relatively static, but as the piece progresses you can (should?) start adding more dynamic elements (modulation, pitch contour, ornamentation, dramatic swells, etc.).

SPATIAL PLACEMENT

Each player should be panned in a shared sound system so that their localization in the sound field corresponds to their position on stage. (If separate speakers for each player are available, all the better!)

MESSAGING

For the "1.0" version, you will need to send and receive these events.

`+ Send`

`note`

`+ Receive` (addressed to you OR `/hub`)

`respond`

`note`

/hub (or /hubster) <from_hubster> respond (for v1.0, <from_hubster> will always be Matt)

`stop`

In your software, make sure there is a way for you to VISUALLY monitor the `respond` message (a Max `button` object or something?). Then design your algorithm to automatically react to this message:

IF you are currently sustaining a `slat` note:

— Turn off the `slat`.

— The software can optionally play a simultaneous short `post` note here (essentially "cutting off" the slat)

ELSE (you aren't currently sustaining a `slat`):

— Turn on your button/light/etc., to display that this message has arrived.

When you do NOTICE your button/light/signal go on, MANUALLY trigger either a `post` or `slat` to play. Try to respond as fast as you can! The idea here is to create interesting rhythmic patterns based on reaction times.

```
/<hubster> <from_hubster>
note <fMaxDelayInSeconds>
<fMinDelayFac> <fCps>
<fTransposeFac>
```

The response to this message is all automatic. Have your software react:

+ CALCULATE the "minimum delay value":

`minDelay = fMinDelayFac * fMaxDelayInSeconds`

+ PICK a delay value (in seconds) that is in the range of your calculated minimum value and `fMaxDelayInSeconds`:

`fMyDelay = (minDelay <-> fMaxDelayInSeconds)`

+ DELAY for fMyDelay seconds

+ PLAY a short `post` note at frequency fCps

+ CALCULATE – the next person's pitch value:

`fNewCps = fCps * fTransposeFac`

+ if (40 > fNewCps < 10,000), then SEND a new `note` message WHERE:

`hubster` is another player

`from_hubster` is you

`fMaxDelayInSeconds` is the `fMyDelay` value you used, above

`fMinDelayFac` is the same `fMinDelayFac` originally passed to you

`fCps` is the `fNewCps` value you just calculated, above

`fTransposeFac` is the same `fTransposeFac` originally passed to you

`/hub matt stop`

Automatically respond:

```
+ turn off all currently
playing sounds
+ kill any "delaying"
notes (without playing
them)
+ simultaneously (and
immediately), play a
short note
```

VERSION 2.0 ADDED FEATURES

+ `/hubster matt sendmode <iMode>`

values for iMode:

0 = no instantiation allowed

1 = you are now allowed to send `note` messages manually

2 = you are now allowed to send `respond` messages manually

3 = you are now allowed to send `note` and `respond` messages manually

+ allow note messages to be sent to `/hub ??`

Mark Trayle became too ill to perform with the Hub in 2013, until his death in 2015. Matt Ingalls replaced him in late 2013. Ingalls is the founder of sfSound, a group that sponsors an annual electroacoustic music festival in San Francisco, in addition to other events.

INCOMPETENT ROCK

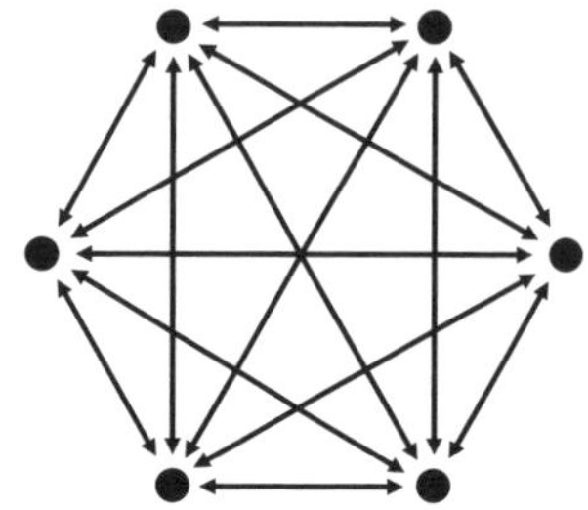

Phil Stone

September 16, 2014

Premiere: October 27, 2014 at Mills College, Oakland, CA

In many ways, this piece is a reiteration of *Borrowing and Stealing,* but is generalized in a way that I hope will explode what that can mean.

Compose/play live/draw/map brainwaves/steal from another Hubster [1] /whistle/ a monophonic musical motif and describe it by a graph of two parameters: [pitch, amplitude] vs. time.

Limit the number of points in the shape to 500. We can negotiate this limit, if necessary.

You may use whatever timing you like, within reason (it's controlled by the delay parameter). It is unlikely that we will get a chance to play hour-long motifs. Many small motifs might be fun. Millisecond-scale-motifs or single-note transformations might also be a good time for all.

Note also that you can specify silence (for instance, for staccato articulation) by using a parameter set with <whatever> for pitch and 0 for amp. When requested by another hubster [2], blast a series of messages back to them describing the frequency/amplitude values that make up the motif [3].

The motif may be played once or many times; individually triggered, looping, or whatever. Taste and discretion, etc. etc.

You should probably have an "all notes off" panic button somewhere in case this piece generates hung notes due to network anomalies or imperfectly-formatted data. (This is probably a good Hub practice in general, right?)

[1] Shapes copied from another Hubster may be used as is, or modified in any way and played. For instance, you may copy both amplitude and frequency, or just one or the other. Note that (as in *Borrowing and Stealing*) this is a loophole for a simplified implementation. If you just want to steal/transform/play, that is perfectly acceptable. It would be good if not everyone did this, though. If you do go this way, try to make some really interesting transformations.

[2] To request a shape dump from another hubster, send this message: /<hubster-name> <your_hub_name> gimme

[3] Motif-description message format: When someone sends you a `gimme` message, respond with:

`/<requesting-hubster_name> <your_hub_name> start <number_of_points>`

Then, for each point of your shape, send a distinct message of the form:

`/<requesting-hubster_name> <your_hub_name> point <point_number>`

(float, `0-9999` frequency in Hz.)

(float, `0.0-1.0` log. amplitude)

(float, delay until next point, in milliseconds)

`<point_number>` starts at `0` and counts up.

Finally, after all points are sent, send:

`/<requesting-hubster_name> <your_hub_name> end`

NOTE that you don't have to send the words `freq`, `amp` or `delay` anymore, just the three values, in freq-amp-delay order.

NOTES FROM SUBSEQUENT EMAILS: One thing we learned about *Incompetent Rock*: try not to blow up if the shape-dump you get from somebody else isn't perfectly well-formed. This is actually pretty easy to do with the new data format Chris suggested which specifies a number for every point in the shape you can use as an index to store that point. Just make sure any "gaps" that appear in a dump (perhaps caused by a network burp, or whatever), don't cause you to blow up (e.g., initialize a background of silence, or random notes, or whatever doesn't crash you, over which you write the incoming shape).

My thinking has changed a little about the "publish what you play" rule. I now think it's perfectly okay to decide what you want to publish, and when. If you take this to mean that it's okay to lie about what you're playing, that's a valid possibility, as long as you can live with yourself. Yet another shortcut to implementation!

One interesting difference from *Borrowing and Stealing* is that the melodic motifs here are not in any specific tuning system or scale—note pitches are defined in Hertz, not by note number ("float" indicates a floating-point numerical value, i.e., a real number). This is made possible by the band's change to software synthesis and is a good example of following the technology's lead.

Tim<

APPRECIATIONS PART I

JULIAN ROHRHUBER
SHELLY KNOTTS
THOMAS LEHN
GEORGE E. LEWIS

THE IRRITAINING MATERIAL OF CAUSALITY

Julian Rohrhuber

1.

It is quite rare, but some suffer from a peculiar condition. One hand suddenly acts on its own, like someone else's hand, turns into a being that turns its will against its owner. This disturbing behavior is called the "anarchic hand syndrome"[1]; imagining all the complicated situations that arise from it, one can imagine that it makes everyday life really difficult for those who have it.

2.

In a softer form, however, the dissociation of bodily actions from conscious control is much more common indeed. In fact, one such case is just in the everyday habituated movement, say of routinely doing the dishes or walking down the stairs. When such an action goes astray, this may be irritating. But it may also be entertaining, or to use a better word irritaining, to realize that some part of ourselves has its own life.[2] For a purely instrumental action it usually counts as something like a convenient shortcut leading towards an otherwise intended goal.

3.

This shortcut of body memory carries dance and instrumental music. But here, the subtle dissociation of body parts may suddenly enter center stage—a hand that is playing on its own, a foot that runs away with the bass drum, a voice that goes beyond the capabilities of voice. In improvised music, the origin of a sound can be intractable, for audience and musicians alike. A sound may be an inadvertent response, it may be an answer to a question nobody asked, an idea from nowhere. It also can be a lucky mistake, or a glitch of an instrument. Also in fully written-out, composed orchestral music, the intricate moiré of timbres is hard to trace. And in forms like algorithmic and aleatoric music, music may become an observatory for events in which the musician is part of the audience.

4.

Many twentieth-century conceptions, and maybe most prominently the *musique concrète* tradition, aims to emancipate sound from its source and enable the listener to have a self-evident and autonomous auditory experience. On the other end of the spectrum, there is also an ear curious for "knowing why," an attention that is directed at following not only the effects but also the causes. I think that the shady zones of contemporary music lend themselves to such a causal hearing.

5.

There are two great reductions, however, which likely de-motivate curiosity about the causal structure of music. The first is the idea of pure decision, embodied in the will of a composer for forming the musical material. If we know that everything that happens is an intended form, then there is, in the end, no cause other than an overall form that is to be carved out by the composition, and in the event of a performance, the individual events like the limbs of a marionette, all hang down from the threads of the composition. Even in aleatoric music, one can detect a negative form of this control, in the idea of events that just happen without a specific reason. In both cases, form is autonomous from its causation. A second reduction is in the importance of the idea of communication, more common in improvised music. One can disassemble the causal relation in a way that each musician is fully autonomous and we hear an exchange, a musical panel discussion. Music speaks, the reason that something was said is to be found in what was said before. In all their richness, these forms encourage all kinds of listening, but they usually do not particularly encourage causal listening.

6.

Let's suppose that among the artistic materials, like sound, paint, wood, and so on, there is the material of causality. Indeed, much of art does invent causal relations, makes things happen in a particular way, intervenes in the state of affairs, composes reasons for why something is. And also here, the material is not a passive matter merely to be formed, but resists, entangles, and demands participation. Causal art is not so much a narration of causes than caused by its very material. Perhaps algorithmic music is the best example? More than just an efficient specification of sound synthesis and musical score, a musician or composer is up to her neck in descriptions of procedures and constraints, in delegation and abstraction, in reasoning about what happens why and why what happens is not what was expected. The music gets out of hand. That's why one may want to listen less for an expression of an intent than for a diffraction of causes within what happens why.

7.

Despite its inner causal abundance, much of computer music gives reason to believe that eventually, all the important stuff happens in a black box on stage. The fact that a performer controls it doesn't help that much; it remains a controlled black box. The League and the Hub made a difference by turning the black box inside out. Network music stages the diffraction of causes. In every piece, it presents a causal network as a specific situation, a specific distribution between persons, programs, protocols, things, and so on. This encourages one to wonder how things happen. There is no first mover and no final cause. Anything may act or intervene, unplanned chains of action may unfold and merge to an event. Borrow it or steal it. Who did what, and who is to be blamed for it? Network music is anarchic in the proper sense of the word: It frees the curiosity for causes from the greed of liability.

8.
In other words, what we have here is a great vehicle to navigate the landscape of being under, in, and out of control. Surely, much of musical practice emancipates actions from the musicians' ability to fully take conscious possession of them, and vice versa. In network music, we encounter technology fallen in trance. Any member, human or not, may become possessed by any other member. The resulting irritainment is more than just enjoyable. It suggests an adequate attitude for a movement beyond the law of the sociotechnical jungle.

1 Johannes Roessler and Naomi Eilan. *Agency and Self-Awareness: Issues in Philosophy and Psychology* (Oxford: Oxford University Press, 2003).

2 The word *irritainment* is borrowed from Laura Kikauka and Gordon Monahan, who used it to describe their ambivalent background music.

LIVE CODING, ALGORAVE, AND THE HUB: THE SOCIAL AND POLITICAL DIMENSION

Shelly Knotts

The live coding scene began around the year 2000 with Alex McLean and Slub's early Perl-script performances in nightclubs, and with the creation of Julian Rohrhuber's JITLib (Just In Time Library) for SuperCollider.[1] The practice, which involves the improvised writing and editing of computer code to generate music or other media, lead to the formation of the not-so-temporary TOPLAP (Temporary Organization for the Promotion of Live Algorithmic Performance) and its associated manifesto that attempted to define parameters of the practice.

As with the Hub's approach to network music, the live-coding manifesto (and subsequent attempts to describe the practice) focuses on the interaction between a performer and an algorithmic system rather than musical style and parameters. TOPLAP (toplap.org) has since become an online repository of information on live coding, as well as a vague organizational principal for the international live-coding community, which forms itself around local "TOPLAP nodes" that organize concerts, meet ups and workshops.

Precursors of live coding include Ron Kuivila's 1985 performance at STEIM using the Forth Music Language,[2] as well as the practices of the Hub that extend and remap algorithmic systems in performance. Beyond technical implementation, the lineage between the Hub and the live-coding scenes can be seen in the social and political dimensions inhabited by both. The Hub's predecessor, the League of Automatic Music Composers, began during the advent of micro-computers, which marked an important moment in the democratization of computing as the (relatively) cheap and portable machines allowed the computer to move from large, funded institutions into the hands of artists and the public realm. John Bischoff recalls how, as the League's networked KIM-1 computers played during biweekly events in Berkeley in 1979, audience members would listen, watch the band tinker with the system, and ask questions.[3]

The algorave scene (algorave.com), a sub-set of live coding, focuses on live-coded dance music. In its engagement with beat-driven music forms, algorave can be credited with moving live coding out of primarily academic and experimental music contexts and towards mainstream music spheres. With algorave, the participation of non-academically affiliated coders exploded, resulting in a diversity of audiences, styles, and new technologies for live coding.

But with its roots in the open-source-software movement, the live-coding scene has always foregrounded transparency, process, and keeping humans in the loop. "Show us your screens" was emphasized in the TOPLAP manifesto and remains central to the live-coding practice today. This is often achieved by projecting the code window onto a screen behind the performer, allowing the audience to follow the process of writing and editing code, respond to the unpredictability of the practice, and share in the moments of risk, error, and failure when inevitable faults creep into the code and music.[4] All of this echoes how the Hub has, on occasion, visualized aspects of their network flow in performance for the audience to see, beginning with their realization of John Cage's *Variations II* in 1995.

In the last decade algorithms have crept into the public consciousness such that they have become synonymous with harmful military, surveillance and big-data technology practices. In algorave, however, audience members are confronted with small and non-harmful algorithms inhabiting arts and social spaces, bringing people together through communal cultural experience. Through witnessing a fallible human interacting with an algorithmic process, code and technology becomes humanized.

Schroeder and Rebelo describe how "live coding practitioners ask the audience to share the risk and the fascination of live making. By emphasizing the risk of such making, these practices deliberately expose the body in flux, the body in constant negotiation with the environment and the instrument, itself in flux."[5] This aesthetic of performing with unpredictable systems and embedding the imperfect human performer within computational feedback loops is inherent in the Hub's work. Tim Perkis's *Waxlips*, for example, defined a form of message-sending that foregrounded emergent structures arising from group interaction. The piece could easily fail if a group member missed a message or if the network was incorrectly connected.

In live coding, however, far from being undesirable, this type of error brings about a rupture in the performance that takes the audience out of the code and back to the performer(s) and the collective human experience taking place. There is an ephemerality in the code, data, and network structures that we can create quickly but that we can also destroy just as quickly.

1 Renick Bell, "Considering Interaction in Live Coding Through a Pragmatic Aesthetic Theory" *Proceedings of SI13, NTU/ADM Symposium on Sound and Interactivity.* Nanyang Technological University, Singapore, 2013.

2 Andrew Sorensen, "Impromptu: An Interactive Programming Environment for Composition and Performance" *Australasian Computer Music Conference 2005: Improvise*, 2005.

3 Chris Brown and John Bischoff, "Indigenous to the Net: Early Network Music Bands in the San Francisco Bay Area," 2002, available online at http://crossfade.walkerart.org/brownbischoff/IndigenoustotheNetPrint.html.

4 Shelly Knotts, "Live Coding and Failure," *The Aesthetics of Imperfection in Music and the Arts: Spontaneity, Flaws and the Unfinished* (2020): 189.

5 Franziska Schroeder and Pedro Rebelo, "The Pontydian Performance: The Performative Layer." *Organised Sound* 14, no. 2 (2009): 134–141.

HOW THEY HUB: GROUP IMPROVISATION AND MUSICAL NETWORKS IN ELECTRONIC AND ACOUSTIC DOMAINS

Thomas Lehn

An important difference between acoustic and electronic instruments is that the former gives the player direct, physical access to its vibrating, sound-producing components. Although electronic instruments—especially those that are digital or software-based—can generate sounds of greater complexity and with multiple, superimposed processes simultaneously, the player does not have the same level of contact with the source of the sound generation, whether they use physical controllers or not. Moreover, loudspeakers are a mandatory part of an electronic instrument's sound production. Yet, when they are used in acoustic ensembles, their sound radiation characteristics often prove disadvantageous for balanced, musical listening, even with optimized position alignment.

What distinguishes many electronic instruments from acoustic ones is their ability to confront the player with self-generated sounds. By navigating between control and letting things happen, electronic performers are challenged to take a position on how and to what extent they should integrate the machine's activities into the overall musical context.

How do these characteristics of electronic instruments and their differences to the acoustic ones affect the contemporary performance practices of improvisation and interpretation? How do they affect the practice of composition and its notation, particularly in regards to repeatability across performances?

In free improvisation, more than one person is simultaneously involved in the act of real-time musical creation. It demands from the performers a constant balancing between individual and collective processes. The ability to immediately access imagined or intended sounds is of the utmost importance in these situations, and therefore it is critical for an electronic musician to find a technical solution that provides the greatest amount of flexibility when playing.

But what if an artist has a completely different set of artistic goals? For example, the Hub is certainly not an improvisation ensemble. But is it an interpreting ensemble? Evidently not in the traditional sense either, since it does not feature classical instruments, nor conventionally notated compositions. Rather, the Hub's six computers are interconnected within a networked system which, depending on the piece's specification, allows each player to generate and share data with the others, often affecting the sounds of the other instruments. The results are unpredictable, and herein lies a parallel of what happens in group improvisation with acoustic instruments: It is mandatory that players spontaneously adapt to what's going on in each moment of a performance.

The compositions of the Hub are foremost characterized in the particular computer programming specified for each piece. Within these frameworks, the players interpret the pieces by navigating their realization in a listening-adaptive manner, improvisationally rendering the compositional ideas embedded in the interaction of their instruments. "We allow the network interaction to make it impossible for us to predict, or exactly control, the behavior of our instruments," Chris Brown explains.[1] "You steer the instrument more than you control it. Part of the idea is to push it this or that way and watch it go, and then try to influence it in various ways."[2]

Seen as *one* instrument, the Hub is entirely different from an acoustic improvising ensemble, yet how it navigates the course of its performances is akin to improvisation. The way improvised music unfolds in acoustic ensembles is perhaps more characterized by immediacy in individual and collective action and the intentionality of each player's spontaneous sound production. However, the collective musical result of such ensemble improvisation cannot be intentional (nor by its nature does it aim to be) since what each of the players will do next is unpredictable. But it is also far from random because each player's actions are connected to the actions of the others—through listening, through meta-levels of perception, and through their direct engagement. The improvisers are intertwined in a network of their ears and minds, and it is within this network that the entire musical progression in group-based free improvisation is formed.

The same could be said about the performance practice of the Hub, but with one major difference: A network of computers, each with its own set of programmed behaviors, is involved and interactively interfering with the network of minds. Could this ever-changing movement of the performers' minds be explained any better than in Alvin Curran's comment: *"Imitate, ignore, dominate, tacit, and be free—this is how we play and how we hub."*?[3]

1 Chris Brown in an email, March 31, 2021.

2 Chris Brown in a Zoom conversation, May 23, 2021.

3 From Alvin Curran's liner notes from the CD *Wreckin' Ball.*

THE HUB'S TELEMATIC SOCIALITIES

George E. Lewis

According to Scot Gresham-Lancaster, one of the Hub's founding members, it was fellow experimental composer Nicolas Collins who "created" the Hub by securing two New York City venues, the Clocktower and Phill Niblock's Experimental Intermedia, for the group's very first performances.

Those now legendary performances were telematic. Three members of the band—John Bischoff, Tim Perkis, and Mark Trayle—were based at one site, Chris Brown, Scot Gresham-Lancaster, and Phil Stone at the other; 2400-baud modems connected the computers, allowing the performers to exchange information for each piece between the two sites.[1]

The English-language term "telematic" is a translation of *télématique*, a merger of the French words *télécommunication* and *informatique* coined in a 1977 report to the French government by Simon Nora and Alain Minc, which was eventually published as *L'Informatisation de la société*. This merger was not merely semantic; Nora and Minc predicted that the convergence of telecommunications and computers would have incalculable but powerful effects on the world. "Slowly but surely," the authors predicted, "telematics will affect the major instruments of culture: language, in its relations to the individual, and even in its social function; and knowledge, as an extension of collective memory."[2]

Expressive culture was hardly exempt from this understanding; by 1984, not long before the creation of the Hub, Roy Ascott was writing that "our emergence into the new world of telematic culture... calls for celebration at the interface to those PDP systems that can link us with super connectivity, mind to mind, into a new planetary community."[3]

In 1998, Gresham-Lancaster observed that "although the group performed at separate locations a few times, it created its strongest and most interesting work with all the participants in the same room, interacting directly with each other and with the emergent algorithmic behavior of each new piece."[4] But telematic processes cannot be so uncomplicatedly defined by geographical distance alone, since in most cases that distance is arbitrary, effectively hidden from both performers and audiences—a kind of "spooky interaction," my pun on Einstein's critique of the theory of quantum entanglement as "spukhafte Fernwirkung" or "spooky action at a distance."

Thus, like quantum entanglement, which can take place at any distance, even in the same room the performers are nonetheless in different spaces, their networking grounded at the nexus of the interface between the humans and the computers. Here, I'd like to suggest that the interplay between two desires—first, to hide distance, and second, to assert it—provides pleasure in telematic encounters, both for direct participants and for audiences. Moreover, the elimination of physical distance as an affective factor in a telematic musical interaction also creates an ironic psychic connection between the immediacy of the experience and the announced fact that the interaction is taking place between entities whose places are ostensibly widely spaced from each other geographically.

As fellow Hub member Perkis has remarked, the Hub's work has concentrated on "using computers to define a new social context for music making, as well as exploring the possibilities for systems too complex for direct control."[5] A more recent version of this is the communication between the Mars Rovers and their handlers on Earth, where due to the fundamental constraint of the speed of light, a radio signal sent between the Rover and Earth will take approximately thirteen minutes to arrive. Thus, the designers "built in lots of autonomy to let the spacecraft take care of itself."[6] In other words, the Rover and its handlers, like the Hub, are telematic improvisors, performing according to the fundamental conditions of improvisation: indeterminacy, agency, analysis of conditions, and choice. This leads me to assert that the telematic musical experience embodied by the Hub is inextricably bound up with two crucial elements: real-time improvisation and the networked agencies of interaction.

And now, because of the recent pandemic, we are all telematic, rendering the vision of the Hub both prescient and urgent.

1 See the heading "Nicolas Collins Creates the Hub," in Scot Gresham-Lancaster, "A Personal Reminiscence on the Roots of Computer Network Music," *Leonardo Music Journal* 27 (2017): 73.

2 Simon Nora and Alain Minc, "Computerizing Society," *Society* 17 (1980): 29.

3 Roy Ascott, "Is There Love in the Telematic Embrace?" in *Telematic Embrace: Visionary Theories of Art, Technology, and Consciousness*, ed. Edward A. Shanken (Berkeley and Los Angeles: University of California Press, 2003), 237.

4 Scot Gresham-Lancaster, "The Aesthetics and History of the Hub: The Effects of Changing Technology on Network Computer Music," *Leonardo Music Journal* 8 (1998): 41.

5 Ibid., 39.

6 Thomas Ormston, "Time Delay Between Earth and Mars," 5 August 2012, https://blogs.esa.int/mex/2012/08/05/time-delay-between-mars-and-earth/.

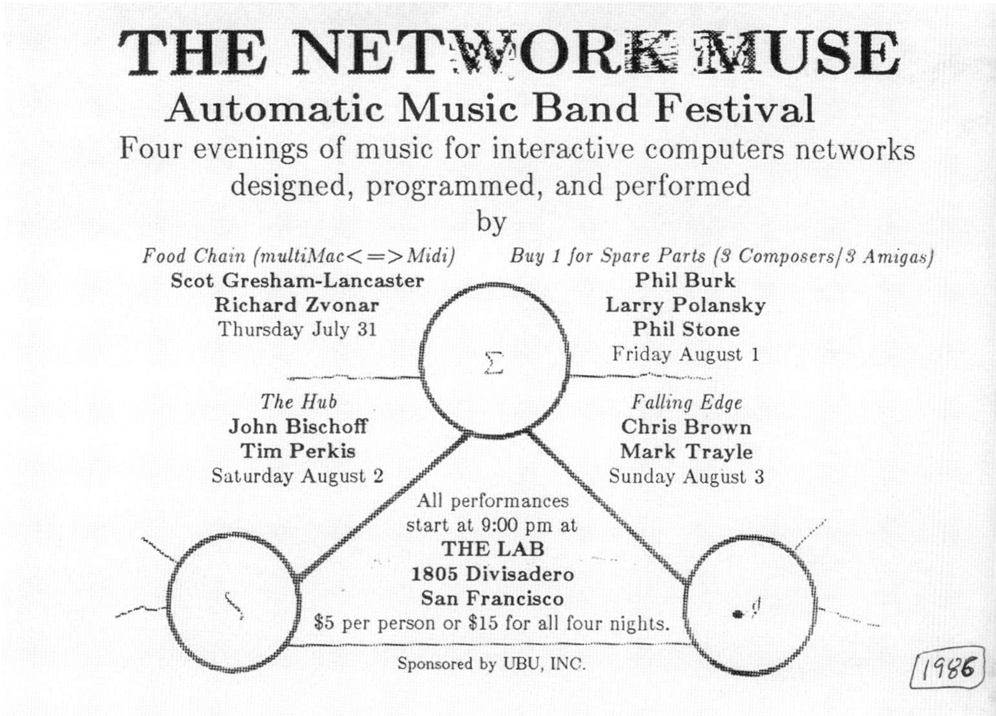

Postcard for the Hub-originating Network Muse Festival, July 31 – August 3, 1986.

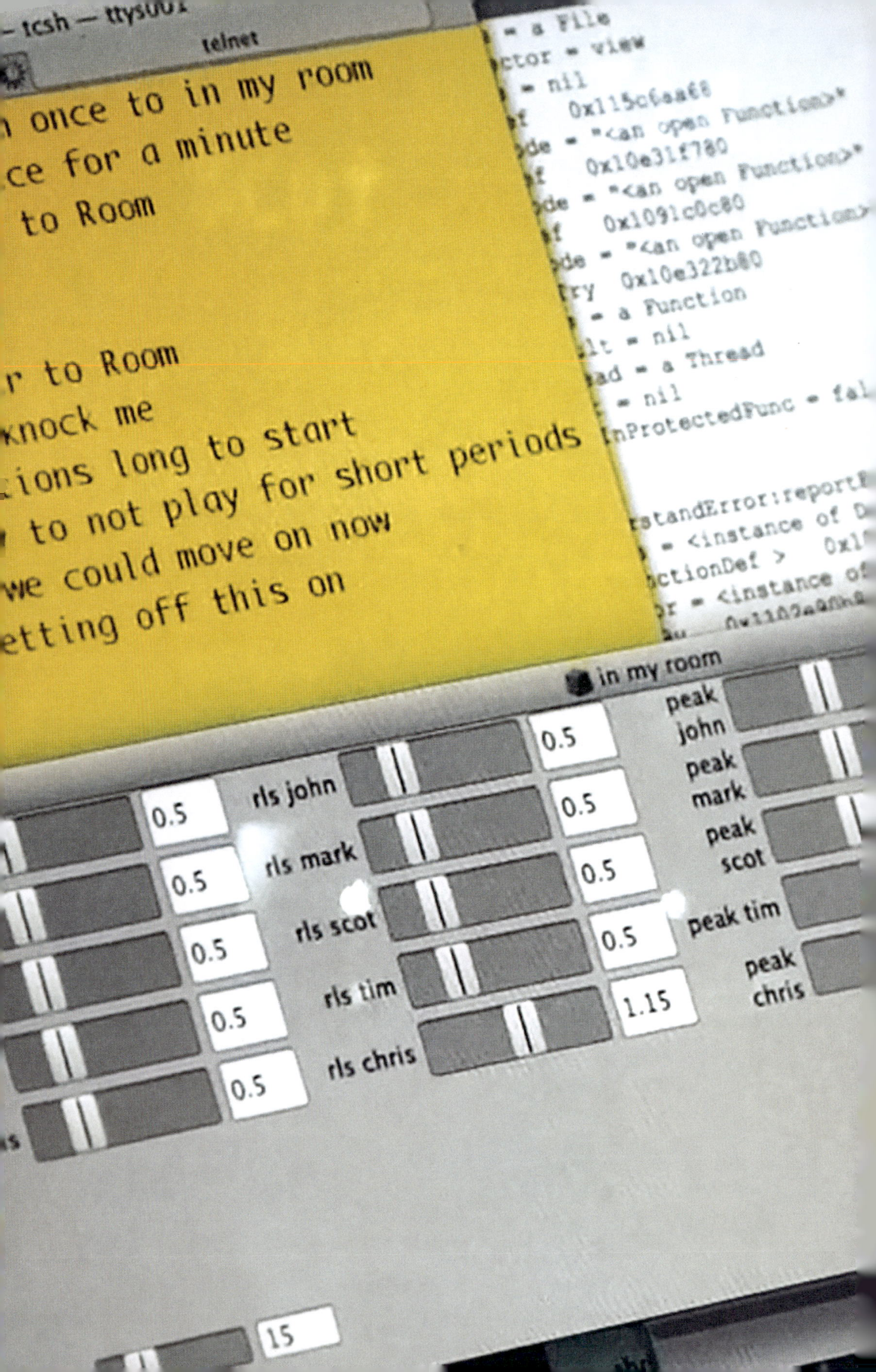
tcsh — ttys001
telnet
once to in my room
ce for a minute
to Room
r to Room
knock me
ions long to start
to not play for short periods
we could move on now
etting off this on
in my room
rls john
rls mark
rls scot
rls tim
rls chris
peak john
peak mark
peak scot
peak tim
peak chris
0.5
1.15
15

VIRTUAL ROUNDTABLE DISCUSSION: THE HUB ON THE HUB

GINO ROBAIR

As the elements of this book began to take shape, it became clear that the members of the Hub should have a chance to speak for themselves about the unique aspects of their practice. I have known each of them since 1986, following the group since its inception, so I was curious not only to learn more about how they have navigated the disparate elements of networked computers and improvisation, but what it was like for each of them to work in this unusually communal, non-hierarchical way for more than three decades.

Rather than have a single roundtable discussion with all members present at once, I felt I would learn more by having individual discussions with each person, beginning with the same set of questions. Approaching the process as a "virtual" roundtable would not only give me additional time to ask follow-up questions but would allow us to explore tangential topics as they appeared. The interviews with John Bischoff, Chris Brown, Scot Gresham-Lancaster, Tim Perkis, and Phil Stone took place in March 2021.

How did the Hub's approach at the beginning differ from what you were doing with the League of Automatic Music Composers?

Bischoff: With the League, the idea was that you come in with your idea of what music is and you play it simultaneously with the other guys. So it was more chaotic in that way. It was as if each player was in their own world, but they were interconnected. The information exchange was a way to pull these disparate things together.

We were all developing our own solo-music ideas separately. Jim Horton was interested in just-intoned melodies and theories of melody, so he was always generating streams of melodic ideas. I was interested in different things: I had one piece based around 8-bit waveform generation, sort of like wave shaping and kind of noisy. I also had a piece that was a bit like a David Berman piece where I listened to the other two League members and analyzed their pitch, but I only played when I detected a harmonic interval and I tried to reinforce that. Tim Perkis would have something completely different that might be a rhythmic idea; some state-changing machine that he had made.

Whereas, in the Hub, even though we were all developing our own music, we adapted whatever we were doing, sound-making-wise, to these unifying specifications. It was quite different in that way and more integrated from the get-go. Less disparate in terms of texture.

How would you compare the networked-ensemble experience between the two groups?

Perkis: It was similar in the sense that what was characteristic about both situations—and so different than any other musical situation that any of us had experienced—is the idea that you are a performer, but not exactly. There were times in the League where we would sit in the front row and listen, then get up and maybe tweak some knobs or move a little cable or something like that. But the idea was that your interaction with the whole system was at some other time level. You are definitely not deciding what every note is. You're not a player like that at all. It was totally new and not like anything else that we knew of at the time.

The idea was that we are composers, we made this thing, and it's acting, and we tweak it a little bit here and there. We just make fine adjustments. It was a way a lot of people in those days used analog synths like the Serge; the idea that there is this great thing happening in a patch that has a dynamic of its own and you're just kind of tweaking it a little. You might speed something up a little bit, and the sound goes a little higher than it did before.

But with the League it was digital and we'd get these weirdly non-linear things. It wasn't smooth like an analog synth. It would jump into some new kind of weird state all the time and in unpredictable ways. That's what was so exciting about it: The idea that you didn't know what it was going to do; the emergence of some complex thing that nobody planned or put in there.

The initial idea for using something like the hub was that it would allow for a more precise, specifiable set of interactions. You write things to certain memory locations and people are supposed to read certain things, so the whole process is more predictable. And all the messages would get to where they were going.

Whereas, the League was really more like free improv; there are ways we are connected, but who knows what they are, and who knows what the effect will be. I might have some kind of serial line [with a voltage] that's going up and down which is connected to Jim Horton, but who knows what happens along the way or what he does with it. So there were ways we were coordinated that you could hear, but it felt crazily unpredictable by comparison.

The device that became the band's namesake was initially a pair of Synertek SYM-1 computers the group used for the event in New York City.

Stone: Yes. For their duo, John and Tim had worked out a design for a new-generation hub: a KIM-1 with a couple of ACIA's connected to it so they could connect it to their serial ports. It had an RS-232 interface, which was standard back then. When we got the gig, Nic Collins suggested we play in two separate spaces simultaneously, and we realized we needed to build a more robust version of that hub. And that's when we came upon the idea of using two SYMs.

Tim did all the hardware work on it; he wire wrapped the boards for the four ACIA's for each one. And I wrote a program that basically made shared memory out of those boxes: It was modeled on what they were using earlier than that but rewritten and optimized. And each Hub constantly copied its shared memory out of the fourth ACIA so that it would automatically write to the other hub. Each hub was copying its memory to the other hub and, therefore, we had six people visible on a hub at one time, if you will. And those two different SYMs—one in each space—were connected by modem: Three of us were located at the Clocktower, and the other three were seven blocks uptown at Phill Niblock's Experimental Intermedia Foundation. The six of us weren't physically in the same space but our data was shared between all six players.

What was the data transfer rate?

Stone: With the SYMs, it was effectively about a kilobit [per second]. We could read and write to the hub at 9600 bits per second, but it took three or four bytes to get the job done. It also took time for a byte changed in the "other" hub to copy over to your local hub. Slow as molasses by modern standards in other words, but fast enough for us. We used that setup for about two or three years.

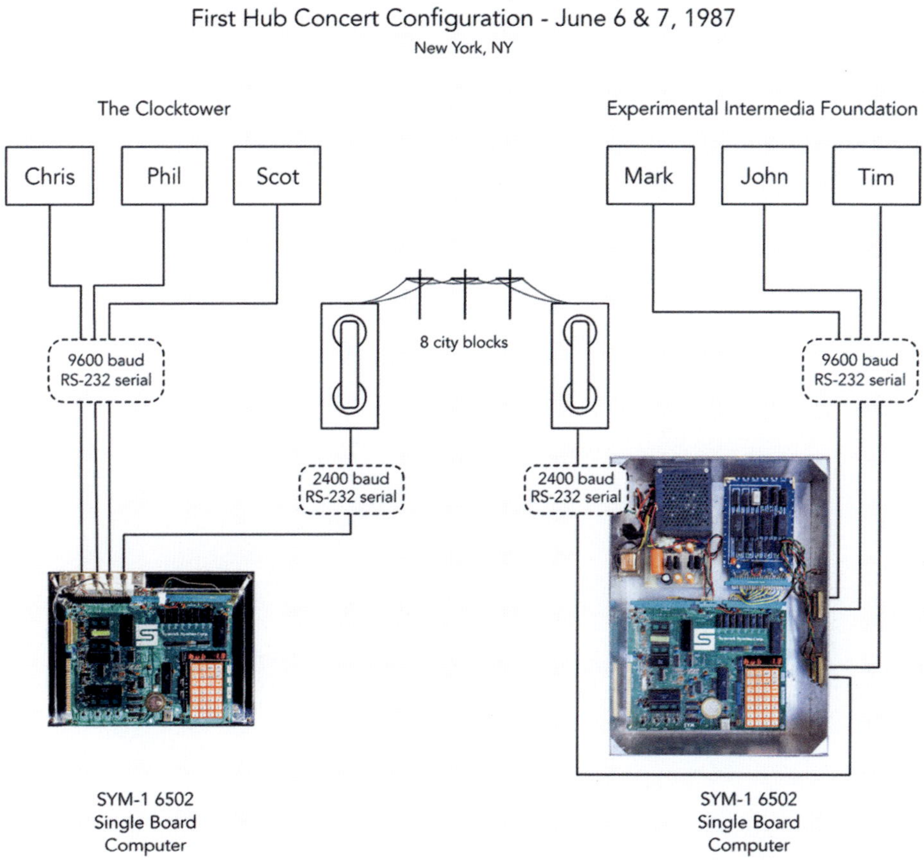

Configuration of the Hub at its first performance, which occurred simultaneously at two locations in New York City, June 6 & 7, 1987. Graphic design: Judy Coates Perez.

Bischoff: Even when we were all in the same space, it wasn't fast enough for rhythmic synchronization. So that was a major limitation. We were sending different parameters for different pieces which didn't require any kind of tight coordination. Because of the lower speed, a lot of those early pieces were forced into a situation where the information exchange was more embedded and diversified within the texture of the group. There was a lot of correlation in the group, but it would be very hard to pinpoint. Whereas, once we got into more rhythmic synchronization with the MIDI protocol, then you could recognize these specifications. You could hear the structure, the idea, the information exchange, and the messy, diversified realization of it. You could actually hear that in the music, whereas in the early stuff it would be very hard to pull it apart.

Chris Brown rehearsing with the Hub at Mills College, Oakland, CA, October 6, 1989. Photo: Jim Block.

What was the next iteration of the hub?

Stone: We started using the Opcode Studio 5 as a MIDI hub, which Scot reprogramed to act as a star network instead of the conductor-orchestra configuration that MIDI is usually set up for.

Gresham-Lancaster: I was a consultant for Opcode at the time and knew Dave Oppenheim, so I was able to get access to the ROM code. He helped me make the changes. Basically, the box resets the MIDI channel of every message going through it to match the input-port number that the MIDI message came in on, and then it routes that message to the MIDI output-port number that matches the message's original MIDI channel. For example, I was hooked up to Port 7 and if I sent a message on MIDI channel 2, it would show up at Port 2 on MIDI channel 7. John was hooked up to Port 2, so when he got a message on MIDI channel 7 he would know it was from me. That is how we identified each other: John was on Port 2, Chris on 3, Tim on 4, Mark on 5, Phil on 6, and I was on 7. Channel 1 was open for everybody—the general-purpose, omni sort of setting—so it would blast out to the whole Hub.

Did using the Studio 5 and MIDI change the Hub's music?

Stone: It definitely changed the nature of the pieces because we could do high-speed, directed communication with each other. We weren't just dropping notes in the mailbox which a person could pick up at their leisure. We could send a piece of information from one Hub member directly to another at about five times the rate of speed, because it was that much faster than the RS-232 we were using. So, it enabled more high-speed types of things to happen.

But as much as we really didn't want to get stuck in the MIDI paradigm of there being 128 notes and 128 degrees of control, the pieces were still influenced by it. More often than not, we were passing MIDI Note numbers around for something. *Waxlips,* is a good example. The idea of using note numbers is baked into MIDI and if you don't avoid it, it becomes a part of the structure.

Gresham-Lancaster: And notes and phrases are not what electro-acoustic music is about. Its strength is in the metamorphosis of sound, the ability to change timbre over time, which wasn't present in the 1983 vision of what a universal music language should be.

But although MIDI notes ended up being a thing instead of textures, we broke that pretty quickly with my piece *Stuck Note*. The title refers to how stuck notes were a problem with MIDI: If you sent a Note On but didn't send a Note Off, the note would keep playing. The spec is very simple. You are allowed to use only two controls on your synthesizer—volume and one other parameter of your choice. At the beginning, you play one note, and that note plays throughout the piece. Everyone but you has control over your volume and x-parameter. So, you can turn everybody else up and down, but you don't know what they're going to do to your instrument.

Brown: We also got more rhythmically tight at that point because we could define the actual meaning of the message that was being sent. That's when I came up with the idea for *Wheelies*.

I had just been to Cuba and was writing polyrhythmic music and wondering why people who were into pulse weren't working with polyrhythm, because it's so much more free.

In *Wheelies*, we all shared a stream of real-time MIDI clock messages that I generated using the shape editors in HMSL. I would draw tempo streams and send them out to the group. The basic idea was that everybody would conform to that stream, which was very fast. But they could make a beat from some subdivision of the stream and share cyclic relationships with other players. That was basically the spec.

But after that, I felt like I didn't want to keep making pieces where I was the conductor. Although I distributed the decision-making freely to the group in *Wheelies*, I was still putting out this pulse stream and shaping it, so it was like I was conducting the beat. And I wanted to get away from that in my next piece.

Stone: When we went to the pure UDP-based network of the modern hub, we got beyond MIDI and now we tend to send frequencies (or whatever else we want) around instead of note numbers.

Gresham-Lancaster: Once OSC arrived, we could just use our names—/Tim or /John—instead of channel numbers.

Perkis: And when we started using OSC and everybody was doing direct synthesis of different kinds, we were less bound by the rhythmic characteristics of a MIDI event. The results were slower moving but with this kind of broader, timbral evolution.

It might be surprising to some that the Hub's pieces are highly performative and require active listening as well as improvisation and interaction with structures.

Stone: That's critical to it. The whole basis of this band is electronic improvisation. And then structure is imposed on that by the specs. That first concert got us noticed because we played in two different places at once. The *Village Voice* did a nice review of it and were kind of amazed by that possibility.

Bischoff: But we felt that playing as a band where three people could only hear one trio at a time was a big drawback. You knew the other people were affecting what you were doing, but you couldn't hear them.

Stone: We didn't really want to continue doing it that way because we like to be in the same room while playing. It's a radical concept, but that's the best thing about the ensemble; hearing each

Phil Stone rehearsing with the Hub at Mills College, Oakland, CA, October 6, 1989. Photo: Jim Block.

other, reacting to it, playing with it. That's the important part; not the fact that digital data is flying around. The end-product is the fun part, and that requires being together, socially.

Has the Hub played in other telematic contexts since that premiere performance?

Stone: Well anytime we've had to do it, it has been a lot less satisfying. And there is a conceit to it: It's like the audience is supposed to be impressed that this is happening; that there is an invisible hand somehow affecting this performance. Conceptual art has its place, but I'm much more interested in the end-result. Like I said, we were just missing half of our bandmates in that

Scot Gresham-Lancaster rehearsing with the Hub at Mills College, Oakland, CA, October 6, 1989. Photo: Jim Block.

situation. I don't know if I speak for everybody, but I think most of us really did not dig the telematics part. It was just an accident of our birth. We did it two other times, and a dozen more times said no: If you want us, we're going to play in the same room.

Which is interesting considering the network aspect of the Hub, especially during the recent pandemic where musicians often played together over Zoom.

Stone: It is a constant struggle to push back against the idea that network music is the same as Internet music. It's not the same at all. I recently wrote a journal paper in which I said: If you want to talk about network music don't make it synonymous with Zoom jams. That's not what it is.

There is definitely a place for Zoom jams: It has kept us interacting during a very tough time and it will continue to be useful. But it really has nothing to do with this idea of musicians interacting digitally while they play live.

Gresham-Lancaster: There *was* the expectation that we would continue doing these kinds of remote performances. But we enjoy being on stage as a group and the phenomenon of using the network to make algorithmic decisions that the network reacts to in unexpected ways. When you're playing in the Hub, a lot of times you are just sort of riding the bucking bronco and don't know if it's going to start off a flurry of notes or just be quiet for a minute. Things just happen, depending on the piece.

And the hub acts as a mediating device?

Gresham-Lancaster: Yeah. The scores have different phenomenology associated with the way they are constructed and many times that is in the score. But a lot of times that is also in the heterogeneous quality of the way we realized that score. For example, *Waxlips* is the most networked piece. It does not work if all of the members of the network aren't doing it. We say that somebody has a *hole*, meaning a note was coming in but it wasn't coming out because there was a programming issue.

John Bischoff rehearsing with the Hub at Mills College, Oakland, CA, October 6, 1989. Photo: Jim Block.

The phenomenon of the piece is made up of these holes. We also came up with the vernacular of saying *hose*: we'd *hose* a bunch of notes in and see how long it would continue. Sometimes the piece would keep going and sometimes it would die out, and then we would hose more notes in. We have always been playing with the bugs of our own code, which were not uniform because everybody was using different software.

That's very different from the current laptop-orchestra concept, where everybody has the same software and the pieces are already defined; and when you get the piece, you run it the way it was designed to run. It's very composed and doesn't have that heterogeneous quality that I think is unique about the Hub.

So every member of the Hub writes his own software interface for each new piece?

Stone: Yes. With a few exceptions, that has been the Hub practice all along: Everybody writes their own response to the spec.

When the Hub first started, how did you approach composing for the ensemble?

Brown: It isn't exactly composing. We call them specs, instead of compositions. You have an idea and how it would be realized, but then it takes its own shape as everybody makes their own version of it. As you play together, you figure out a way for it to actually work. And because the music is made with computer-based technology, it has to be improvised. There isn't a notation system or a way to make a score and follow it. Improvisation is inevitably part of it. And particularly, when I joined the Hub, I felt like that was the main way I could participate because I was not as good at coding as everybody else was. I hadn't worked with it as long, and certainly not making live music with computers. That was the part that was a challenge for me at the beginning.

For that reason, the first piece I made for the group, *Rol'em*, was entirely based on improvisation and was an attempt to get the music out of being just with computers. I knew all of us played acoustic instruments, and I was responding to what I was used to when collaborating with other improvisers, which is to make up the structures that could use the technology but weren't really based on something that technology could do that humans couldn't.

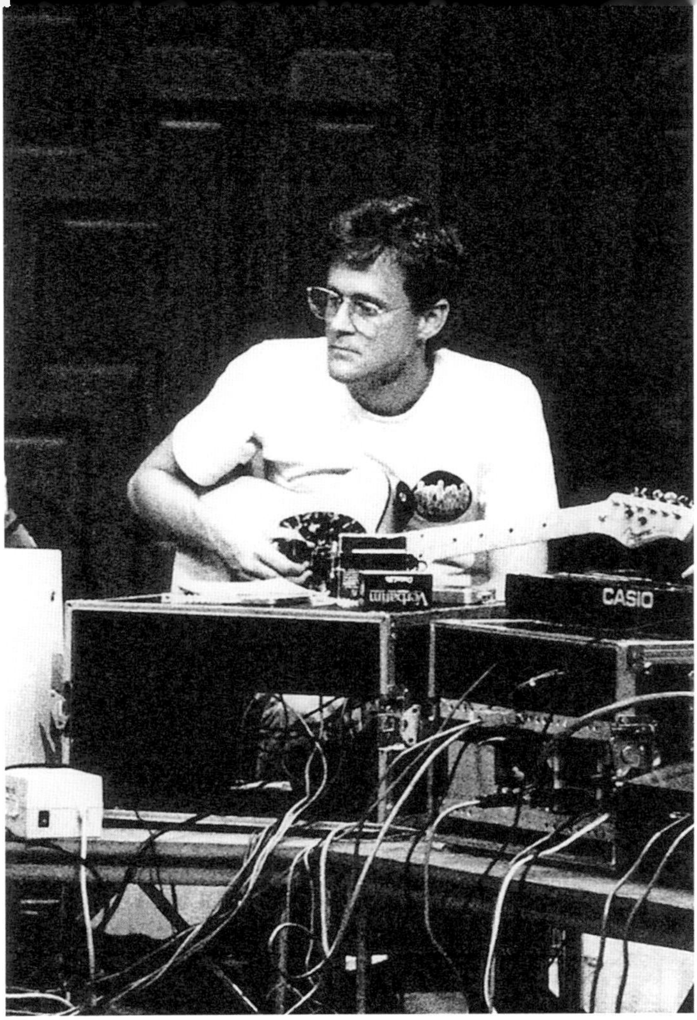

Mark Trayle rehearsing with the Hub at Mills College, Oakland, CA, October 6, 1989. Photo: Jim Block.

How did you structure that piece?

Brown: In a lot of ways it was inspired by playing Christian Wolff's music, where there are explicit cues between the players and the music is organized around them. The computer makes choices about who gets what cues and what scenes were being asked for by random choice through the computer. It would have been difficult to throw dice or use some other sort of chance operation to choose between a number of possible scenarios, so I described subsets of the group—trios, quartets, quintets, sextets. At the beginning of every new section, the computer would decide what the next ensemble would be. It's like the computer is the prompter with a chart of the various possibilities that describe different kinds of improvising ensembles.

One of the things that makes it stand out in our repertoire is that it specifically asks people to play acoustically, even to the

Tim Perkis rehearsing with the Hub at Mills College, Oakland, CA, October 6, 1989. Photo: Jim Block.

extent they make sounds by playing the keyboard box or physical things that are related to the computer. So, the computers are making sounds that aren't always electronic. They're just being struck in the same way that you might strike the body of a piano or a cello.

And I should add that composing for the Hub probably wasn't always the major part of being in the group. The major part was always *realizing* pieces or ideas that people would have.

Stone: *Borrowing and Stealing* was the first spec that I did for the ensemble. We had to get that ready for the New York concert, for which we were severely under-rehearsed. We were so busy getting the technology working in time for the gig that we were lucky to get those pieces together with the amount of time we had left.

It is a fairly comprehensive spec because, basically, it was also a tutorial on how to work with this new box that we had built; how to approach its address space. *Borrowing and Stealing* divided the address space up into six parts so we each had our own place to store the melody we were playing. And then if you wanted to copy someone else's, you just went over to their space and read it back into your system.

Bischoff: It's also good piece for describing how those early hub boxes worked because you would report whatever melody you were playing into this shared memory, and you could choose to play your own melodies, or you could play other people's melodies, or any combination or transformation in between.

Often, my specs were influenced by my solo work. I would just think about what kind of sound do I want to hear, what kind of texture, and what have we not done. I tried to push things in new directions, as I did in *The Glass Hand*. Almost all the work we had been doing up until then was with short sounds, or sounds where there was space in the music, and I thought: "Let's do a piece that is just dense, noisy textures—all at once." And so I built a spec around that. I think it was fairly successful and certainly made the group sound different.

When I look at Mark's pieces, they're beautifully structured and great examples of a kind of way that the Hub was successful. Their charm is that you can hear the spec in them; it is very clear that the spec is this structural thing. That's a constant, and it highlights the diversity of response, the messiness of the group response to that. And if those two things together balance each other, it's amazing, really.

Stone: Mark had a sense of finding the essence in what was needed to make a network piece. And that's not to take anything away from the type of work the rest of us have done, because that has produced amazing pieces. You put in all that work on a well-designed spec, it's going to come out good. But it's so fun when a simple spec generates complex behavior.

My favorite one to use as an example for that is *Waxlips* because that is a bone-simple spec. Yet it produces what Tim calls emergent behavior; behavior that is not expected, but it's there somehow in the essence of the system.

Perkis: For me, the trouble with most of the Hub pieces was that everything was too complicated. I felt like we had too much randomness to know whether there was anything going on in terms of real emergent behavior. I wanted to do an experiment to see if we could reduce the situation and gain some clarity and hear these transformations. *Waxlips* was an attempt to try to really strip it down and say: "How simple can it be and have it *only* be this interaction; the only thing that generates anything? How much can you limit it and have it not feel boring and stiff and not alive?"

So the idea is that you set up a rigid set of rules about what you do. And there is no randomness involved: You have no random choice as a player as to what you play, or some kind of independent, algorithmic decision making. There is just a simple, fixed mapping. You create a table for yourself, and you just do the same thing every time you get the same message from the same person: If I get a request for a B-flat from John, that means I always just play that, and then I ask Scot to play an F-sharp or whatever. So the same input gives exactly the same output for a section. Each section of the piece has a fixed behavior. Then, for the next section, I send a message to everyone to use a new map.

What's interesting is that it gets into these weird, chaotic sequence-states. When some message comes in and starts this chain of reactions, it's a predictable reaction that will always happen with the same messages, yet it still had a lot of variation because the timing is never exactly right, and the messages are all asynchronous: I get something and it takes a certain amount of time for me to send this thing to you, and then it goes through the MIDI hub box that delays it a certain amount of time. And there is more than one of these chains going on at once. So the piece found a kind of a sweet spot in terms of how complex things got and how intelligible things were.

And when you change the section and mapping, the emergent behavior is recognizable in a different way. It has a certain integrity.

Perkis: Right. There will be a moment where I send the message that says "make new tables" and then, bam; the music would change immediately and sound quite different.

It was very hard to debug because if somebody's setup is not working, there is a leak and all the notes just run out, and the whole thing stops. I realized that every now and then I had to spew some messages to seed the whole thing to keep it going.

It was the only piece that had normal music-sounding things that just arose out of the interaction that nobody put in there. Especially in version 2, [*Waxlips II* (*Das Lied von der Nerd*)] where I also limited the pitch set. It wasn't just a full chromatic set of pitches, but you adopted a new scale, or new pitch-set, when you re-mapped things.

I think it was successful as a Hub piece. It was intelligible bu still really interesting and unpredictable. You could hear what we were doing and what the mechanism was.

The Hub performing at Mills College, Oakland, CA, February 13, 2016 (L–R) John Bischoff, Scot Gresham-Lancaster, Phil Stone, Matt Ingalls, Tim Perkis. Photo: Phil Althouse.

Has the way you conceive of specs changed over time? Not just because of the changes in technology but in the way you work as a group?

Gresham-Lancaster: It has changed a lot because when we were first making definitions, we didn't know how hard they would be to implement. You can see this in the definitions as we went along, and the way that, as pieces were put together, they would become much more simplified and easier to realize.

My piece *Listeners Digest* was very sophisticated and used just intonation up to 11-limit intervals and then down to a 3:2 tuning ratio. And we worked and worked on that, and got a version that worked. But ultimately, when we did it on tour, I did it as a solo performance because it was just too much work for the others on top of all the other work they had to do.

So that was a real lesson for me personally: I'm going to make things that require one night of coding. I know everybody's coding level. The nice thing about the Hub is that all of us are aware of the amount of work we're asking the others to do and we're doing that amount of work ourselves, so there is a kind of equity that way.

Each of you understands where the others are coming from.

Gresham-Lancaster: Yeah, right. And we know our weaknesses. It is ongoing. And the technology is such a moving target. That changes everything, as well.

I was also interested in getting the insider/outsider perspective of the Hub's newest member, Matt Ingalls. Although a generation younger than the others, Ingalls was an established composer, improviser, and computer-music programmer when he was asked to substitute for Mark Trayle in late 2013. He has continued playing with the group since Trayle's death in 2015. I spoke with Ingalls in June 2021.

How did your involvement with the Hub begin?

Ingalls: I had the idea that sfSound could try to play some Hub scores with just acoustic instruments—as a chamber ensemble without the Hub members. They were skeptical that it would work. At some point in our negotiations, it turned into a Hub concert and I would be filling in for Mark because he was unavailable. But I still had that idea in the back of my mind because some of my own music for acoustic instruments had these sorts of algorithmic elements. So once I joined, I realized an old piece of mine could be ported to a Hub spec.

It is interesting to see how different your spec for *Fences* is from other Hub pieces.

Ingalls: It definitely seemed different than what they normally do. For one, I wanted everyone to play out of their own speaker, or at least have their own location in the stereo field. And that, according to some members, is off-brand for the Hub. That was actually one of the things I always found frustrating as an audience member; that you couldn't tell who was doing what, especially with six people playing electronics. That's a lot, particularly as technology got better and everyone could do more stuff in real time.

The other thing about the piece is that I was interested in reaction time, like in a Christian Wolff piece. So one of the main things in the spec is that you had to have some sort of manual button. You couldn't use the network to just have your code respon to a message. You had to have your code respond to the message, then alert you in the physical space somehow, and then you had to physically start your note so that we would get these strange rhythmic responses based on your reaction time. I also specified something about the nature of the sound: When you respond to *this* type message it needs to be some sort of a percussive sound. That didn't quite work. I don't think they were used to doing that.

What was it like stepping into an established group?

Ingalls: There's a lot of history there, and all this unspoken stuff they already know. As the new member, there was so much technical groundwork I had to do for every piece. And it's not like normal music. You get to the rehearsals and you are still trying to figure out all the networking stuff and if the other messages are working. First, you have to debug your setup and sometimes that eats up half the rehearsal. But you're dealing with six people, all of whom have showed up to the rehearsal in various stages of their implementation, and all have probably some bugs and inconsistencies.

The musical aspect always comes at the end, after you've dealt with all the technical issues. I think that's why a lot of their compositions have evolved the way they have. They talk about that a lot; that they're improvisers and everything developed to where the pieces that worked best are more about defining the network and leaving the musical decisions up to individual players.

How would you describe the experience, overall?

Ingalls: Being on the performing side of it is definitely different than being in the audience. The technical aspects are interesting, but it was a lot of work just to get up to speed with that one piece, to lay down all that foundation in my system. Then, with each new piece you have to put in a lot of effort just to get it to work. But it did inspire me to do more networking music, which I'm working on with my colleague from sfSound, composer-improviser Chris Burns.

One thing that I felt could be improved was the way Hub pieces always pass all messages to every performer, which I think goes back to the old hub they used to have. From my work as a programmer, I am a little more familiar with client-server networking and I thought it would be interesting to have *that* kind of paradigm, where you send messages to the server and the server routes it. So I could say "I want to send a message to Gino," and the server actually knows where you are and only sends you that message. In the Hub, basically, you're throwing everything at the hub, and the hub sends it all out to everybody else. When I got Chris Burns involved, we realized you could actually incorporate the server into some of the pieces, so the individual players would have less they have to do. The composer of the piece would write code on the server side that handles some of the algorithmic stuff, depending on the piece, perhaps posting a table of notes or whatever.

So the server acts as a player as well as a mediator.

Ingalls: Yeah. You can write a plug-in for the server that is a piece, but you don't have to. In its default mode, the server only routes messages.

APPRECIATIONS PART II

ALVIN CURRAN
BARBARA GOLDEN
LAETITIA SONAMI
ATAU TANAKA

LOOK FO

PERFORMANCE MUSIC DANCE READINGS

THE HUB

This group of electronic music composer/performers gets its name from an electronic linkage and memory controller (The Hub) which allows the six of them to send and receive their musical information in a performance setting. The audience hears the collective output of the group. Kyle Gann of the Village Voice said "for the listener the exitement lay in the reemergence...of music as process: the original province of minimalism, long since abandoned...were I to place bets on history, I could see The Hub as the beginning of a very important movement".

November 6 & 7
8:30 pm
Friday & Saturday
$5/$4

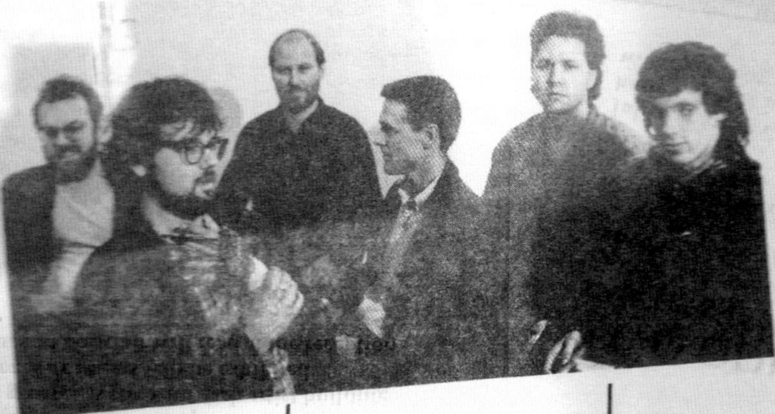

KATHR

Planet Dog

Precisely artic
in this interdi
improvise text
documentatio
Moira Roth, a
Planet Dog "t
years".

December 4 &
8:30 pm
Friday & Satur
$5/$4

IG GRADY/KEITH BAREFOOT

o without Tears

ithout Apples

ngeles composer and musician who designs and builds his struments, modifies existing ones, and from these s, achieves what High Performance magazine calls "an e exoticism, a Middle Eastern and Javanese fantasy put Los Angeles urban guerilla tactics", Kraig Grady will the musical accompaniment to the films *Embryo without* nd *Eve without Apples*, produced in collaboration with ographer Keith Barefoot. In these films, a floating fish, a asement, an amazon with bow and arrow, and a severed eveal the surreal story of a mysterious investigator who e a man from outer space.

the L

OF HUBS HUBBING AND THE HUB

Alvin Curran

Hubs are everyday places of exchange where stuff (e.g., aircraft, trains, trucks, ships, goods, garbage, extraterrestrials, information, and people) arrive and depart for other destinations, often changing roles and uses. Inspired by this structural concept, an electronic music group was founded in Oakland, California in the mid-1980s: It called itself the Hub.

I'm told by informed sources that much earlier, say around the Mesozoic Era, the big animals of the plains, coming from far and wide, were known to meet sing and dance in certain locations. Reliable witnesses say that the animals themselves coined the word *hubbing* for these events, because the movements seen from above appeared as lines of animals moving en masse at various speeds, converging on an imaginary central point, and then moving back outwards, at various speeds, in somewhat more random patterns from whence they came.

From my childhood, more remote Hub fantasies come to mind: the hubcap—the automobile wheel covers, often abandoned on the sides of roads, which in the '60s were thought to be the origins of the common plastic Frisbee, a sports object apparently used for connecting several people by their tossing a circular plastic disk between them.

In the beginning the Hub founders were aware of these precedents, both evolutionary, structural and conceptual.

When I arrived in Oakland to occupy the Milhaud Chair at Mills College in 1990, two of the Mills professors (Chris Brown and John Bischoff) were already members of the Hub, which then had six people in it. A totally simpatico bunch, dedicated to the unknown. We quickly became friends and then collaborators... my computer ignorance notwithstanding.

The collective spirit of participation in real-time performance, man-machine-chance-operations, instant-composition, and acceptance of failure and ecstatic success as equivalent outcomes, are in fact the common modes of ego suppression one enters into while Hubbing or making any music which requires stepping outside of oneself.

It is here that we confront and accept behaviors, systems, codes, and other things beyond human control as musical agents of good feelings. Hubbing is not only about absolute trust between humans who do things because they can, or because they do not have to demonstrate any outcome; it is also about sharing in acts of idealized experimentalism as sources of happiness and joy and that these acts along with their unknown outcomes may produce some satisfying musical experience for everyone, players and listeners alike.

I am improvising, trying to give you what I remember of my experience with the Hub, which was all at once mystifying, illuminating, technologically compelling, even potentially annoying, as if on a drum-roll, the band hit a B-flat chord and from the wings came five one-armed jugglers who were pitching a ball to themselves and the others for no reason at all, like in many modern sports… no one questioned why; the answer was in the doing; trusting chance, the highly structured codes, and the hard-soft wares of the machines themselves—dumb, faithful, indifferent, prone to malfunction and failure as they often demonstrated. They left you no choice, other than fly with no wings, trust the concept, relinquish the ego, believe in the bits bytes buts and bats. Hit "start" with love and skepticism; what comes out will always go up and down, produce highs and lows, long and shorts, do's and don'ts, fats and thins, plinks, plucks, pebbles and ponds, nodes and knots, walls and doors, drones and typhoons, feathers and bricks, funnels, sand and loam, puddles, pickles, overtones, standing waves, noodles, glissandi, and clusters.

At this point, I put my wetsuit on, thinking about a high waving portamento about to burst onto dry land that arches and comes crashing down on an early Hub CD full of suggestive titles and words like: *Waxlips* upon receiving the request Das Lied von der Nerd the player starts this section at his own discretion, waits a computed amount of time in Mob Ecstasy near Wave Junction when Tim's Orgasmatron doing a surf-riff of SIMPLE DEGRADATION drizzles code while the Minister of Pitch takes his Gazamba, his Axe Thing, to open Perry Mason's jar of an unstable Glass Hand comb-filtered through a non-composer of collective ego.

So let's go Hubbing already! If you get my data transition rate?

What this can mean to someone like myself who has never once said a Hail Mary to electricity… even while having had some deadly volts pass through his body: I am not the subject here. It is the Hub, which following the late '50s–'60s exciting thrust toward spontaneous music making, realized ways of creating instant group music by subjecting multiple brainless computers to simple sets of human-like commands that would result in a collective composition/performance. No big deal at the end of the twentieth century; on the other hand a very big deal indeed, positing a trusting alliance between humans and machines as well as between humans and humans, all within a shared belief that there exist musical forms that need not demonstrate predictable outcomes, nor bear semblance to what people commonly call music. In my view the Hub's history—in toto—is a landmark map of major sites to visit in our coming musical future.

A SNAPSHOT IN TIME

Barbara Golden

The Hub did not just spring up out of nowhere. We watched it happen and we participated in it, fueled by many nurturing elements. Northern California in the '80s: mild, rainy winters; calla lilies and lemons blooming in February; palm trees, towering redwoods; the palpable thrill of sex, rivers of vodka and tequila; smoldering soldering irons and joints.

Historically all the elements were in place. The San Francisco Bay Area, crazy beautiful, was a youngster metropolis, just over a century away from Gold Rush days, whcrc brothcl owners and roughneck miners became pillars of society. Ensuing decades retained the Maltese Falcon image: louche, foggy, illicit. City Lights Bookstore, home of the Beat Generation, has howled since the '50s. The '60s saw the emergence of the Black Panthers, the Free Speech Movement, and the SF Tape Music Center. Iconoclasm was the name of the game.

The birth years of the Hub guys spanned the '50s: John was born in '49, Tim '51, Chris '53, Scot '54, Mark '55, and Phil in '59. They all played acoustic instruments as children, except Phil who started while in uni. John and Phil worked with the brilliant Paul DeMarinis. John adds that Jim Horton forced him to start building microcomputers in 1977. Tim studied videography at Ann Arbor, got into computer graphics, then made computer sounds for his vids. Chris, who hails Gordon Mumma as a major influence, created contraptions intriguingly named Gazamba and Hot Lunch. In high school, Phil studied BASIC, probably Mark did, too. Scot, the wunderkind, built theremins.

Their "aha" moments were attributed to John Cage, Merce Cunningham, David Tudor, Harry Partch, and "Blue" Gene Tyranny, who dragged a 14-year-old Scot to Mills College where Robert Ashley and David Behrman split the teaching year. The Mills College Center for Contemporary Music (CCM) in Oakland became a major hub (no pun intended) for electroacoustic music in the USA; the new home of the SF Tape Music Center. I think of it as the Darmstadt of California.

They are composers and musicians first, deeply steeped in all kinds of music. And it's *all* about the music, which is tasty, beguiling, jazzy. I don't know why but the nerdy computer boys and I (the ex-housewife schoolteacher from Montreal with brains, boobs, culinary chops and a dirty mouth) gravitated towards one another. I knew sweet noth-

The Band SPAM circa 1982 (L-R) Mark Traylee, Phil Stone, Barbara Golden, Mary Oliver, Carla Fabrizio, and Magdalen Luecke. Photo: Doug Sandberg.

ing about computers, still don't, but had a taste for great music and art. I also threw tons of parties, a salon where people could eat, drink, and relax.

In 1985 Charles Amirkhanian gave me a midnight slot on Pacifica Radio's KPFA that I called "Crack o' Dawn." Mark was my board operator until he "got bored" and I had to do it myself, until 2020 when I settled in Nova Scotia.

I have no idea when any of them ever slept. They all had day jobs. John and Chris both taught in San Francisco. Chris also tuned pianos. Scot was tech director at the CCM, and Phil was a tech assistant for a while. Tim, Phil, and Mark worked as software engineers, the latter two at NASA and TRW. After their daily grind, they all worked on music; their devotion was mind boggling. Some of them even had the stamina to party! It helped that we were all younger forty years ago. There was *no internet*, which gave us all a lot more time than we seem to have now. We also had a kind of "Dr. Feelgood" in the gang, a shy electronic composer who helped some of us stay awake until the wee hours. Hey! It was the '80s!

The early Hub performances were jittery high occurrences. The instruments would stall, freeze, seize, but we, the audience, were indoctrinated cognoscenti. We'd yell out to the guys during lulls. We knew we were participating in history. It was like the birth of the Model T and Marconi radio transmissions. Christ, we were at frickin' Kitty Hawk!

Snapshots:
Mark and Phil were in my first band, SPAM.

First hot tub I was ever in was at John and Tim's fancy rental in the Berkeley hills.

First time I laid on eyes on Scot, he was on the floor of the CCM, surrounded by machines. He introduced me to my third and final ex-husband, Bill Thibault.

Chris's wife, Johanna Poethig, and I formed WIGband, a trash political duo, in 1985 (still going). Occasionally, Chris, Scot, and Mark played backup and did tech. They were known as The Hogs. Phil corrected false facts in my songs: "Barb, the Scullys do not have a Pause button."

From my journal: *4 September 1989. Scot phoned. Ramon Sender wants me to sing (no, read words) for Hub broadcast on Thursday the 7th before I do "Crack o' Dawn." Kenny (Atchley) is my guest.*

We did a 90-minute live event called *Hub Renga*, after a Japanese form of poetry. Writers from all over, via the WELL, could type lines of poetry which we'd read and would immediatcly influence the band's sounds. There is an excerpt on *Wreckin' Ball*. At the very end of the track I utter: "Static." I shall forever wonder which one of the guys decided to give me the last word.

WHEN DID I STOP NOT LIKING THE HUB?

Laetitia Sonami

There is no special date when I realized how essential the Hub was. It was more akin to recognizing a shape thru the fog or getting to know a wild animal... It takes time sometimes to hear (how could I not hear then?).

Here six (white) men untethered by matter and worldly concerns of virtuosity (assuming the classical idea of virtuosity as embodied), let alone mere expressivity, were connected to each other and the gods with a secret language not accessible to mere mortals, winks here, winks there... this was my battlefield.

What about the music you say? I will get to it, but in the '80s some of us grappled with the transition from analog to digital media. Gestures, expressivity, embodiment, let alone the connection to space and audience, were of concern. How to convey causality, how to "connect" even if our presence, voice and gestures were cryptic?

The Hub continued the ensemble music tradition, where corporeality had been slowly relinquished to "popular" music, cerebral and untarnished by moving bodies. Sitting at their computers, their communication was taking place in a linguistic space.

Would it have been different if I saw them sweating while typing, ecstatically pressing the letter *A*? It would have been terribly disappointing. We all witnessed the bending over in pain of the performer-typist. (I do admit, though, being enthralled by some laptop performances such as from Helmut Schäfer in Beijing, which ended up with his laptop flying off stage, unplugging all audio.)

Would it have been different if they dressed up? John Bischoff mentioned that he once wondered about it and put on a tie but was told he looked like an office worker. With or without ties, it was easy to associate them with stockbrokers intently staring at their screens.

What about visual renderings? The attempts at displaying their messages ended up being a distraction, a reductive proposition. Such theatre or artifice would have been out of place in the Hub.

Would it have been different if I was witnessing six women? Let's imagine Laurie Spiegel, Frankie Mann, and four clones? Some of us are simple that way. I probably would have been more intrigued and challenged. Mirror neurons triggered, I would have felt part of it… it would not have made the music better or worse, it would have allowed some connection in the mostly male landscape of "experimental art music" of the time. Is identification necessary for appreciation?, you may ask. This would be a very reductive proposition, but as a female performer, I could not dissociate the social and cultural contexts from the "music itself."

As the history of electronic music gets drawn, it is necessary to understand the values at play and how these defined the gross traits of its performative manifestation. Values mostly drawn by institutions which were the depository of equipment and research and which established how "meaningful" the music was, thus relegating to the background other expressions such as the immersive architectural performances of Maryanne Amacher, Éliane Radigue, Maggi Payne, the ambiguous electronic performances of Paul DeMarinis or Jim Pomeroy, or Michel Waisvisz's expressive gestural performances.

The Hub did not need access. They could build their own systems; most of us did not know how. The Hub did not have to acknowledge an audience. They were conversing amongst themselves and we were witnesses, not participants.

I am not holding the Hub responsible for the exclusion I experienced at the time. Its members are an exceptional gathering of musicians, hackers, instrument builders, improvisers, and thinkers. I am trying to understand what was behind my reluctance to acknowledge their importance and to "hear" their music.

In the last ten years or so, the landscape changed. I let go of my angst for "meaningful" performative gestures. Contours between genres and practices softened, access widened and allowed for more diversity. The recognition that experimentalism was not limited to one particular form but had taken place concurrently in other genres and countries grew. There was not just one club. The canon widened and freed of my prejudices, I (re)discovered the Hub.

The theatre still remains quaint, but the lack of artifice appears now so refreshing. There is no veneer to contextualize or interpret the work, just the music.

With the now predominance of prechewed software creating deep musical ruts, listening to the Hub is exhilarating, the music demented and punk. For me, the Hub is deeply American, and more specifically Californian—playful, ironic, subversive, irreverent, unbridled, raunchy, homegrown, and unapologetic.

The apparently endless resources we currently have in the digital-analog market, with infinite number of modules, buttons, and increasing CPUs defining the "instrument," is an often-neglected proposition. By instrument I mean a system with rules and elastic boundaries which become audible as they are extended and subverted. Each composition played by the Hub is an instrument. Crystalline structures resulting from each player's or each piece's set of rules appear when one listens. The machine speaks, its boundaries constantly redistributed, become "palpable." Code becomes a skin, etched and stretched, and oddly sensual. As AI spreads in musical systems, with intent and adaptation at its core, listening to the Hub is liberating. There is no message and no place to go. There is no future, there is just *now*.

SIGNALS AND HUMAN NETWORKS: THE HUB AT CAGEFEST, SAN FRANCISCO, 1989

Atau Tanaka

Analog modular synthesizers are networks. The patch cables interconnecting oscillators, filters, ramp generators, and timers create rhizomatic, material "diagrams" that carry signal and music. This was my understanding of networks as a budding electronic musician in the late 1980s. I had trained in the analog studio at Harvard on a Buchla 200 and Serge modular with Serge's brother, Ivan Tcherepnin. Ivan had been through the San Francisco Tape Music Center in the 1960s and knew John Chowning, so he encouraged me to apply to Stanford University's CCRMA.

Upon arrival, I discovered that the heart of CCRMA was the Foonly mainframe computer and Samson Box synthesizer. Networks, then, meant accessing the mainframe via the Telnet protocol, the client/server architecture with which we composed computer music, out of real time. But I wanted to go live.

The live music scene was in San Francisco where the venues, clubs, and arts were amazingly diverse. As a New Englander, California was like another country, perhaps as unfamiliar as Europe, but in a different way. You'd drive up I-280 from Palo Alto to Stanford knowing that legends like Neil Young were up in the foothills, and walk around Haight-Ashbury imagining what it must have been like when the Grateful Dead were there. Musicians from New York's downtown scene such as John Zorn came through San Francisco to work with musicians such as Mike Patton. Networks were also people; not just the professional networks but a musical promiscuity of style and exploration.

Turning form and format upside down was what enthralled me about the Bay Area scene. Finding a way to do so while extending "classical music" traditions of ensemble, composition, score, and performance practices seemed to be imminently possible with the still-new concept of live computer music. The Hub embodied that vision. They were a band, an ensemble. Musical networks meant signals, packets, eye contact—*network computer chamber music*. Imagining the future still existed back then, and in pre-Internet-era San Francisco, future, past, and present had glorious collisions.

September 16, 1989: Fort Mason, a decommissioned military base turned cultural center; CageFest, in the presence of the honored composer. Randall Packer produced Cage's *Musicircus*—a musical network of its own. His *Imagi-*

nary Landscapes, whether for conch shell, gong, rubbish bin (No. 2), a dozen radios (No. 4), or three-and-a-half dozen vinyl records (No. 5) are networks. Compositions. Form and format upside down, in the tradition.

It was this setting in which I saw the Hub in concert, a super-group of musicians I'd known about separately. People told me I needed to meet John Bischoff, so going to the concert was my way to check out what he did musically. He was at Mills College, which had its distinct tradition: While CCRMA may have been at the foundations of computer music, Mills represented twentieth-century American music.

I met Tim Perkis and Scot Gresham-Lancaster after the concert and became friends with them over time. My archives of our early communications are lost, but these guys, and the whole of the Hub were heroes to me who, with the compression of the decades, would become artist-colleagues. As I later worked with the ZKM and SWR2 on a network *hörspiel*, I evoked the Hub in my writings.[1,2] I stayed in touch with Scot and crossed paths with him at Locus Solus in Aix-en-Provence exploring sonification and the mobile phone as itinerant, networked music-making instrument.

What was the network with which they made their computer chamber music a "band"? The CageFest gig was not your typical Hub concert. The line-up did not include Phil Stone, but substituted Richard Zvonar. And they were playing Cage's *Fontana Mix*, a graphical score that was itself a network of indeterminate elements. No longer using their KIM computers, the Hub interpreted the score with their pre-laptop, personal computers networked through an Opcode Studio 5 MIDI interface. The blend of rigorous composition and freewheeling improvisation (what we today call "comprovisation"[3]) reconciled my composition degree with playing guitar in bands. Gathered around the network hub, each member with their computer, the Hub served not only as a precursor to the laptop music that would arrive a decade later but was, for me, a form of chamber music where ensemble performance is a networked activity—analog or digital, electronic or acoustic, human or machine.

1 Atau Tanaka, "Composing as a Function of Infrastructure." *Surface Tension: Problematics of Site*. eds. Ken Erlich, Ken and Brandon LaBelle (Los Angeles: Errant Bodies Press, 2003).

2 —, "Sensor-Based Musical Instruments and Interactive Music." *The Oxford Handbook of Computer Music*. ed. R. Dean (Oxford: Oxford University Press, 2009), 233–257.

3 Richard Dudas, "'Comprovisation': The Various Facets of Composed Improvisation within Interactive Performance Systems." *Leonardo Music Journal*. Vol. 20, (2020): 29–31.

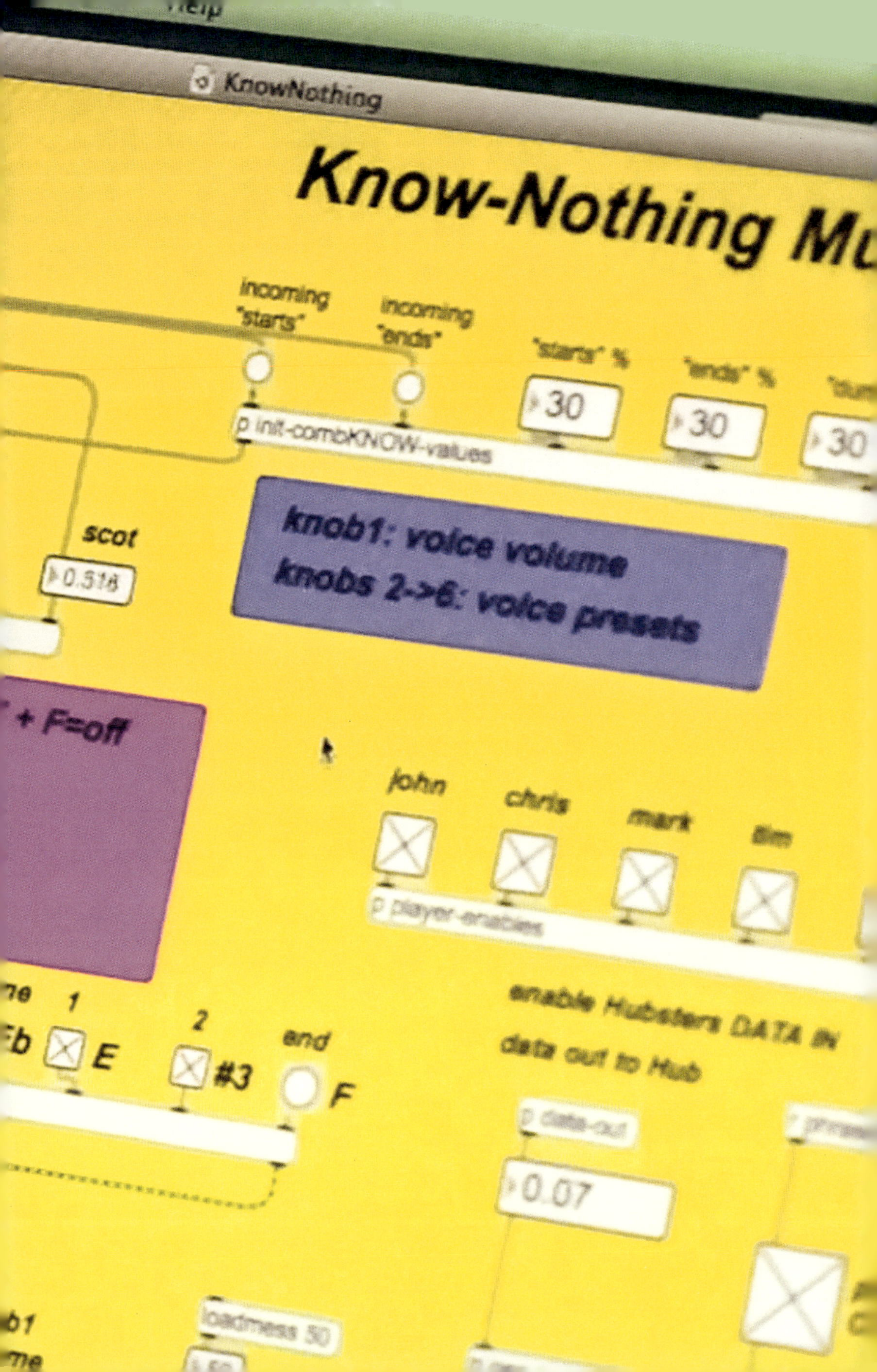

KnowNothing
Know-Nothing Mu
incoming "starts"
incoming "ends"
"starts" %
"ends" %
30
30
30
p init-combKNOW-values
scot
0.518
knob1: voice volume
knobs 2->6: voice presets
+ F=off
john
chris
mark
tim
p player-enables
1
2
end
E
#3
F
enable Hubsters DATA IN
data out to Hub
p data-out
0.07
loadmess 50

John Bischoff is known for his solo constructions in real-time synthesis. His performances around the U.S. include two New Music America festivals, Experimental Intermedia in New York, and the Lampo Series in Chicago. He has also performed in Europe at the Festival d'Automne in Paris, Akademie der Künste in Berlin, and Fylkingen in Stockholm. In 1999 Bischoff was awarded an Artist Grant from the Foundation for Contemporary Arts. He taught for many years at Mills College and was on staff at the legendary Center for Contemporary Music. His music is available on Artifact Recordings, Lovely Music, Tzadik, and New World Records. (johnbischoff.com)

Chris Brown is a composer, pianist, and electronic musician who makes music with self-designed sonic systems that include acoustic and electro-acoustic instruments, interactive software, computer networks, microtonal tunings, and improvisation. His compositions are designs for performances in which people bring to life musical structures embedded in scores, instruments, and machines. Brown has recorded improvisations with Pauline Oliveros, Fred Frith, the ROVA Sax Quartet, William Winant, and Frank Gratkowski. His music is available on New World Records, Tzadik, F'oc'sle, and Artifact Recordings. (cbmuse.com)

Scot Gresham-Lancaster is a composer, performer, instrument builder, and educator working at the boundary between science and art, specifically developing advanced techniques in sonification. He does research and performance using the expanding capabilities of computer networks to create new environments for cross-discipline expression. His interest is in the behavior of interconnected music machines to create innovative ways for performers and computers to interact. Gresham-Lancaster has collaborated in "co-located" performances in real time over networks. He has worked with multimedia prototyping and user interface theory. (scot.greshamlancaster.com)

In addition to being a sought-after composer, computer-music programmer, and clarinetist, **Matt Ingalls** is the founder of the experimental-music ensemble sfSound and co-director of the San Francisco Tape Music Festival. The list of music software he has developed includes Soundflower, a universally acclaimed Mac OS audio utility, the virtual-analog synth Aardvark for iOS, and MacCSound, a real-time version for Macintosh computers of the Csound software-synthesis environment. (mattingalls.com)

Tim Perkis is a well-known figure in the worlds of improvised and electronic music. Over the decades he has played his unique computer-based electronic instruments with hundreds of musicians, including many of the leading players in free improvisation from Europe and the U.S. Along with the Hub, Perkis is a founding member of several electronic music ensembles including FuzzyBunny and Splendor Generator. Recordings of his music are available on the Artifact, Tzadik, New World and Emanem labels, among others. His documentary film *Noisy People* (2007) and the Noisy People podcast can be found at perkis.com.

Phil Stone has been fooling with computers since 1976, either musically or as a way to make a living. He studied experimental music at Wesleyan University and Mills College and was a founding member of the Hub. His programming career includes five years with NASA and 28 years with the University of California. Stone has collaborated for 37 years with the Laura Pawel Dance Company of New York, playing live electronic music. He wrote a *4'33"* app for the John Cage Trust in 2013, enjoys playing electric bass in rock bands, and occasionally builds a piece of fine wood furniture. (pkstonemusic.com)

Ludger Brümmer is currently professor of composition with digital media at the State University of Music Trossingen, head of the Hertz Lab at the ZKM Karlsruhe, and a member of the prestigious Academy of the Arts Berlin. He holds a masters in pedagogy and in composition. Brümmer works in the field of sonic physical modelling, algorithmic composition, spatial audio, and interdisciplinary art. Many of his works try to connect electronic music to historic sources in music like works by Gesualdo da Venosa or Maurice Ravel through composed granulated structures. In this context he works with mathematical models such as fractals and cellular automata. He composed the music of the Amazonas Oper for the Munich Biennale where he worked together with Peter Weibel (text and direction) and Bernd Linterman (visuals).

Alvin Curran has realized a long and fruitful career as a composer/performer, installation artist, writer, and teacher in the American experimental music tradition. His music, whether chamber works, radio-art, large-scale environmental theater, solo performance, or as founding member of the Musica Elettronica Viva group, embraces all sounds, all spaces, and all people. (alvincurran.com)

Golo Föllmer is Professor of Music and Media at the University of Halle. Research on audio art, experimental and electroacoustic music, radio, and acoustic media. Ph.D. 2002 on networked music practices. Projects 2010–2014 on Online Radio and 2013–2016 on Transnational Radio. Curatorial collaboration at sonambiente 1996 and Dystopie Sound Art Festival 2018, among others. Author of radio programs on music and sound, radio play projects with children, audio exhibits, as well as radio and sound art.

Barbara Golden is an artist living in Nova Scotia. The escaped housewife/schoolteacher left Montreal for Mills College in 1979 to study with Terry Riley and Bob Ashley. Since 1985 she has done trash feminist cabaret, WIG band, played Balinese music in the U.S. and Bali, and produced, hosted, and board-operated the acclaimed contemporary music show Crack o' Dawn, on Pacifica KPFA. Golden plays keyboards with The Golden Path, and has performed with Matmos, the Residents, the Hub, and Blechdom From Blechdom. She is currently developing a lounge act for the dive-bar circuit on Nova Scotia's South Shore. (barbaramindagolden.com)

Shelly Knotts is a composer, performer, and improvisor of live electronic, live coded and network music based in England. In 2018 she completed a Ph.D. in Live Computer Music at Durham University, supervised by Nick Collins and Peter Manning. She performs internationally, often using live coding techniques, with a range of styles including Noise, Drone, and Algorave. (shellyknotts.wordpress.com)

Pianist and synthesizer performer **Thomas Lehn** studied sound engineering in Detmold and piano in Köln. He is known for his long-time work as a performer of composed and improvised music with both the piano and the analog synthesizer. Rooted equally in these two major practices of contemporary music, Lehn has developed a unique language of electronic music with his EMS Synthi A. In addition to his own electronic music and extensive international ensemble activities, he has realized compositions by É. Radigue, B. Schaeffer, R. Haubenstock-Ramati, Z. Karkowski, A. Pateras, Chr. Fox, and P. Jakober, among others. His work is documented on about 100 recordings. (thomaslehn.com)

Born in Washington D.C. in 1970, and residing in California since 1983, **Jon Leidecker** is a self-taught composer and improviser, with a long-standing interest in collaboration, collage, and mechanical sounds—or any music which seeks to express more than the self. His lectures on electronic music have been presented at Mills, Stanford, Oxford, UC Berkeley, Peabody Conservatory, and the Museum of Contemporary Art in Barcelona, who commissioned his nine-hour podcast on the history of sampling in music, "Variations." (detritus.net/wobbly)

George E. Lewis is the Edwin H. Case Professor of American Music at Columbia University, where he is Area Chair in Composition and Faculty in Historical Musicology. A pioneer of interactive programs that improvise in concert with human musicians, Lewis's many publications on technology include "Too Many Notes: Computers, Complexity and Culture in *Voyager*" (*Leonardo Music Journal*, 2000) and "Why Do We Want Our Computers to Improvise?" (*Oxford Handbook of Algorithmic Music*, 2018). Lewis holds honorary doctorates from the University of Edinburgh, New College of Florida, and Harvard University.

Gino Robair explores sound using percussion, found objects, and portable electronics. He has composed for film, television, theater, and dance; recorded with Tom Waits, Anthony Braxton, and Lou Harrison; and performed with Nina Hagen, John Zorn, and Wadada Leo Smith. As a music-technology journalist for over 20 years, Robair was editor-in-chief of *Electronic Musician* and *Keyboard* magazines, and he has authored two books. (ginorobair.com)

Julian Rohrhuber teaches music informatics and epistemic media at the Robert Schumann Hochschule Düsseldorf. His works feature algorithmic methods, improvised interventions, and art as a form of thought. As an academic, he writes on topics such as the philosophy of computing and mathematics, sound in documentary film, media theory, and theories of abstraction.

Carter Scholz is a writer, composer, pianist, and improvisor. His books include *Palimpsests*, *Radiance*, *Kafka Americana*, *The Amount to Carry*, and *Gypsy*. His computer/electronic music has been released by Frog Peak and Artifact Recordings, among others. He lives in Berkeley, California.

Laetitia Sonami is a sound artist, performer and composer whose performances, live-film collaborations, and sound installations focus on issues of presence and participation. Best known for her unique instrument, the elbow-length lady's glove, she currently performs with her latest instrument, the Spring Spyre, based on the application of neural networks to real-time audio synthesis. Sonami currently teaches at the Center for Contemporary Music, Mills College. (sonami.net)

Atau Tanaka studied electronic music with Ivan Tcherepnin at Harvard University and computer music composition with John Chowning at Stanford University's CCRMA. He was active in the San Francisco music tech scene, being one of the first musicians to perform with the BioMuse physiological interface. He left the Bay Area in 1992 to conduct research at IRCAM. Tanaka has performed network-music concerts starting at Sonar Festival 1995, and was commissioned by SWR2/ZKM for *Prométhée Numérique*, hörspiel for radio+Internet, 2003. He is currently Professor of Media Computing at Goldsmiths University of London. (ataut.net)

Peter Weibel studied literature, medicine, logic, philosophy, and film in Paris and Vienna. He became a central figure in European media art on account of his various activities as artist, media theorist, curator, and as a nomad between art and science. Since 1999, Peter Weibel has been Chairman and CEO of the ZKM | Center for Art and Media Karlsruhe and, since 2017, director of the Peter Weibel Research Institute for Digital Cultures at the University of Applied Arts Vienna. (peter-weibel.at)

Supplementary Captions

p. 2: Mark Trayle's screen, Birmingham, UK, February 23, 2013.

pp. 4/5: The Hub with Gordon Monahan before performance at Mills College, Oakland, CA, October 6, 1989; (L–R) Scot Gresham-Lancaster, Chris Brown, Mark Trayle, Gordon Monahan, John Bischoff, Phil Stone, Tim Perkis. Photo: Jim Block.

pp. 6/7: Original schematic of the hub box designed and drawn by Tim Perkis, ca. 1986.

pp. 8/9: The band Zero Chat Chat performing at Hatley-Martin Gallery in San Francisco on January 12, 1988; (L–R) Mark Trayle, John Bischoff, Tim Perkis.

pp. 10/11: The "glass" hub, one of the two hubs built in 1987 by Tim Perkis. Photo: Scott Fischbein.

p. 41: At the San Francisco Art Institute in 1987; Scot Gresham-Lancaster wearing a prototype hand controller. Photos: Johanna Poethig.

pp. 48/49: The Hub; dress rehearsal at Mills College, Oakland, CA, October 6, 1989; (L–R) Tim Perkis, Phil Stone, Chris Brown, Scot Gresham-Lancaster, Mark Trayle, John Bischoff. Photo: Jim Block.

pp. 60/61: The Hub; dress rehearsal at Mills College, Oakland, CA, October 6, 1989; (L–R) Tim Perks, Phil Stone, Chris Brown, Scot Gresham-Lancaster, Mark Trayle, John Bischoff. Photo: Jim Block.

p. 67: Mark Trayle, el. guitar (L) and Chris Brown, gazamba (R) performing as Falling Edge, Network Muse Festival, San Francisco, August 3, 1986. Photo: Johanna Poethig.

p. 70: Program for the first Hub performance, which occurred simultaneously at two locations in New York City, June 6 – 7, 1987.

p. 73: The original communication spec for the hub box, 1986.

p. 74: Hub-produced postcard publicizing their November 1987 concerts at The Lab in San Francisco.

pp. 142/143: Tim Perkis' piece *Once* screenshot of John Bischoff's computer, Birmingham UK, February 23, 2013.

p. 148: Scot Gresham-Lancaster's screen with routing overlay display, Birmingham UK, February 23, 2013.

p. 160: Mark Trayle's piece *In My Room*, screenshot of Chris Brown's computer, Birmingham UK, February 23, 2013.

p. 181: (L–R) Scot Gresham-Lancaster and Mark Trayle, rehearsing with the Hub at Mills College, Oakland, CA, October 6, 1989. Photo: Jim Block.

p. 182: John Bischoff rehearsing with the Hub at Mills College, Oakland, CA, October 6, 1989. Photo: Jim Block.

p. 184: A portion of the fall 1987 flyer of The Lab in San Francisco promoting concerts by the Hub on November 6 & 7.

p. 194: Chris Brown's piece *Know-Nothing Music*, screenshot of John Bischoff's computer, Birmingham UK, February 23, 2013.

p. 199: The Hub accepting the 2018 Giga-Hertz-Preis, ZKM | Karlsruhe, DE, November 24, 2018; (L–R) John Bischoff, Chris Brown, Tim Perkis, Golo Föllmer, Scot Gresham-Lancaster, Phil Stone, Ludger Brümmer. Photo: Judy Coates Perez.

pp. 200/201: Hub members outside Tesla in Berlin with Alexanderplatz tower in background, June 21–25, 2005; (L–R) Scot Gresham-Lancaster, Chris Brown, Phil Stone, Mark Trayle, Tim Perkis, John Bischoff. Photo: Jens Brand.

pp. 202/203: The Hub performing at the Roy and Edna Disney/CalArts Theatre (REDCAT), Los Angeles, CA, April 26, 2008; (L–R) Scot Gresham-Lancaster, Chris Brown, John Bischoff, Tim Perkis, Phil Stone, Mark Trayle. Photo: Phil Althouse.

pp. 204/205: Mark Trayle's screen, Birmingham, UK, February 23, 2013.

pp. 206/207: Phil Stone's screen for *Incompetent Rock*, Center for New Music, San Francisco, CA, June 16, 2019. Photo: Judy Coates Perez.

p. 208: The "silver" Hub, one of the two Hubs built in 1987 by Tim Perkis. Photo: Scott Fischbein.

This book was initiated and produced by the ZKM | Hertz-Lab in the framework of the project EASTN-DC, co-financed by the funding program 'Creative Europe' of the European Union.

Editor: Ludger Brümmer (Head of ZKM | Hertz-Lab)
Co-Editors: Cecilia Preiß, Gino Robair
Project Management: Kehrer Verlag (Sylvia Ballhause)
Production Consulting: Yannick Hofmann
(Deputy head of ZKM | Hertz-Lab),
Jens Lutz (Head of ZKM | Publications)
Design: Kehrer Design (Tobias Becker)
Diagrams: Judy Coates Perez
Website: Sabine Jäger, Volker Sommerfeld
Proofreading: Kehrer Verlag (Tom Grace)
Image Processing: Kehrer Design (René Henoch)
Production Management: Kehrer Design (Tom Streicher)

Many thanks to:
ZKM | Videostudio and Christina Zartmann,
Sabine Krause and Dominique Theise,
Sophie Hesse, and everybody who shared their memories and photographs.

Addition thanks from the Hub to Phill Niblock, Veniero Rizzardi, and Hans Koch.

Many thanks to the members of the Hub for co-developing this book.

Bibliographic information published by the Deutsche Nationalbibliothek: The Deutsche Nationalbibliothek lists this publication in the Deutsche Nationalbibliografie; detailed bibliographic data is available on the Internet at http://dnb.dnb.de.

Printed and bound in Germany
ISBN 978-3-96900-041-0

Kehrer Verlag Heidelberg
www.kehrerverlag.com

It is planned to make this book open-access available in the future via the ZKM homepage www.zkm.de/thehub.

This publication has been funded with support from the European Union. It reflects the views only of the authors and the EU cannot be held responsible for any use which may be made of the information contained herein.

CEO and chairman ZKM: Peter Weibel
COO ZKM: Helga Huskamp

ZKM | Center for Art and Media Karlsruhe
Lorenzstraße 19
76135 Karlsruhe, Germany
info@zkm.de
www.zkm.de

Founders of the ZKM:

Baden-Württemberg
MINISTERIUM FÜR WISSENSCHAFT, FORSCHUNG UND KUNST

Karlsruhe

Partner of the ZKM:

EnBW

Supported by:

Co-funded by the Creative Europe Programme of the European Union

European Art Science Technology Network for Digital Creativity

GIGA-
HERTZ-
PREIS
FÜR ELEKTRONISCHE MUSIK
2018

VGS

T: K
C: THE ZEN BEACHBALL
C: OK
SGL: READY FOR
C: LIFE IS A
C: WET NOODLE
J: POINTILISM
M: FLOPPY
C: POINTY
C: MILTON
C: IT'S EVEN
T: ACME SYNTHESIZER
C: WOW, JUST LIKE
M: NOTE PER NOTE
C: DONE
J: COSYNCHRO.
SGL: THAT'S ALL
SGL:
SGL: OKAY NOW HOT
C: HOT TAMALE
J: READY FOR HOT
C: READY
P: READY
T:
T: TATOR TIME
SGL: OKAY I AM READY
SGL:
T: WE SHOULD START
C: I'M BUSY BACKING
Keio Uni

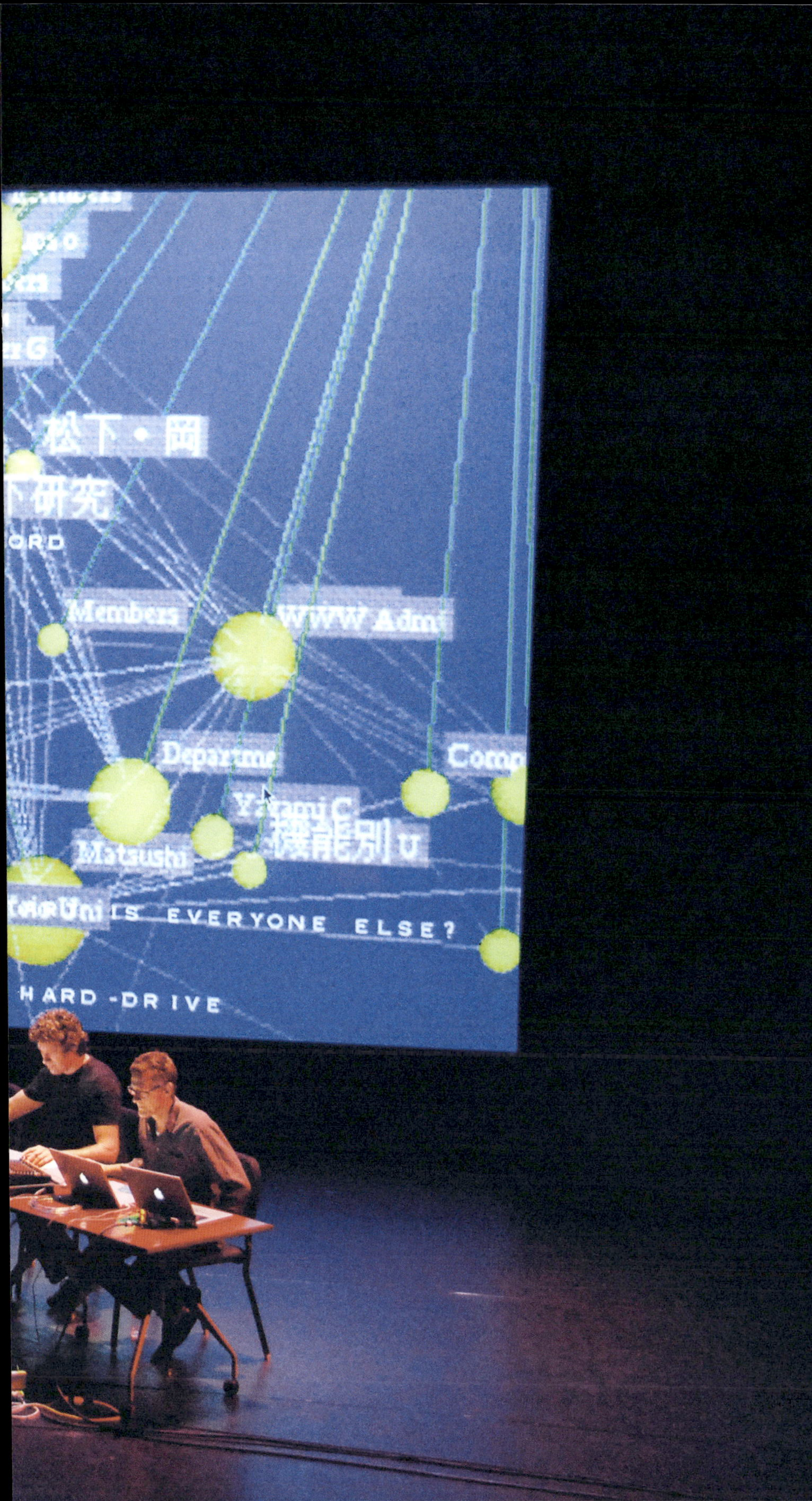
松下・岡
Members
Matsushi
機能別
IS EVERYONE ELSE?
HARD-DRIVE

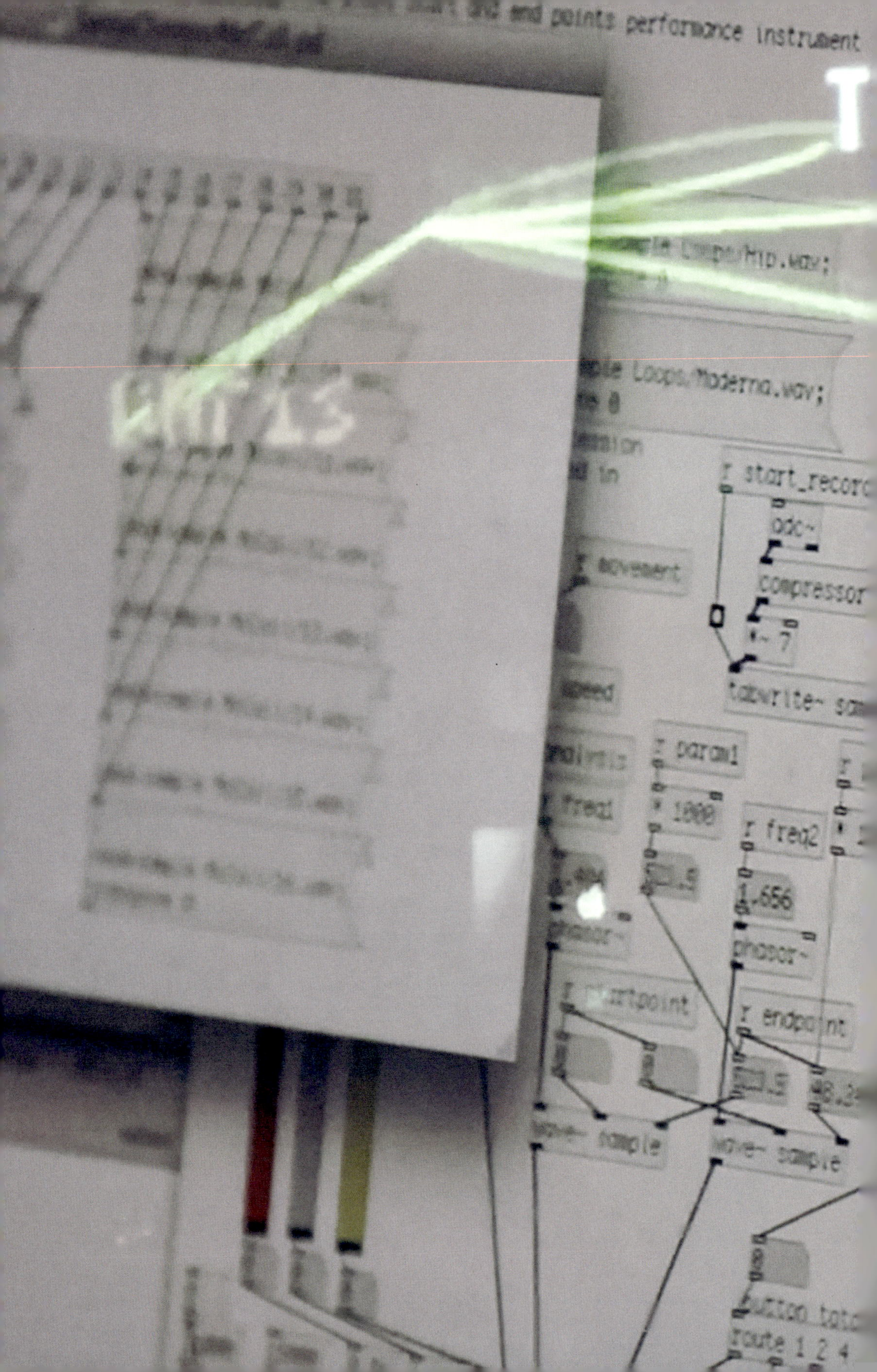

nstrument is for use with
chording to multiplies the controls and does granular
pansion and contraction
ndard playback.
John
d-sample
pd insample
pd phase-tables
d hanning-window
frequency
insamplength
toscale 200 5000
ndpoint
hit print and configure device number
- 0.5
s freq2

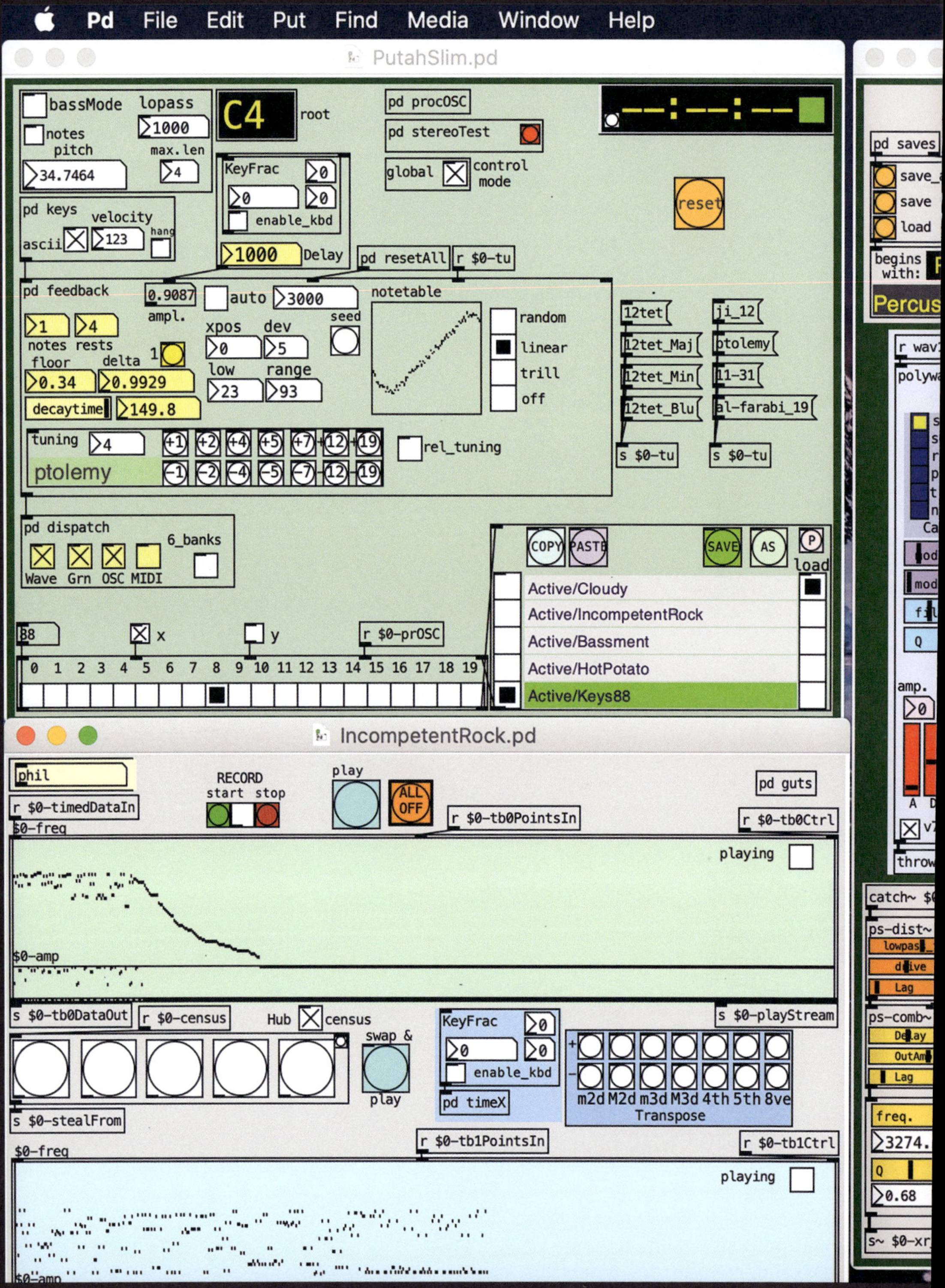
Pd File Edit Put Find Media Window Help
PutahSlim.pd
bassMode
lopass
notes
pitch
max.len
C4
root
pd procOSC
pd stereoTest
global
control mode
KeyFrac
enable_kbd
reset
pd keys
velocity
ascii
hang
Delay
pd resetAll
r $0-tu
pd feedback
0.9087
auto
3000
notetable
ampl.
xpos
dev
seed
random
linear
trill
off
notes
rests
floor
delta
low
range
decaytime
149.8
tuning
ptolemy
rel_tuning
12tet
12tet_Maj
12tet_Min
12tet_Blu
ji_12
ptolemy
11-31
al-farabi_19
s $0-tu
pd dispatch
6_banks
Wave Grn OSC MIDI
COPY
PASTE
SAVE
AS
load
r $0-prOSC
Active/Cloudy
Active/IncompetentRock
Active/Bassment
Active/HotPotato
Active/Keys88
pd saves
save
load
begins with:
IncompetentRock.pd
phil
RECORD
start stop
play
ALL OFF
pd guts
r $0-timedDataIn
r $0-tb0PointsIn
r $0-tb0Ctrl
$0-freq
playing
$0-amp
s $0-tb0DataOut
r $0-census
Hub
census
s $0-playStream
swap &
play
KeyFrac
enable_kbd
pd timeX
m2d M2d m3d M3d 4th 5th 8ve
Transpose
s $0-stealFrom
r $0-tb1PointsIn
r $0-tb1Ctrl
$0-freq
playing
$0-amp
catch~
ps-dist~
ps-comb~
freq.
3274.
0.68

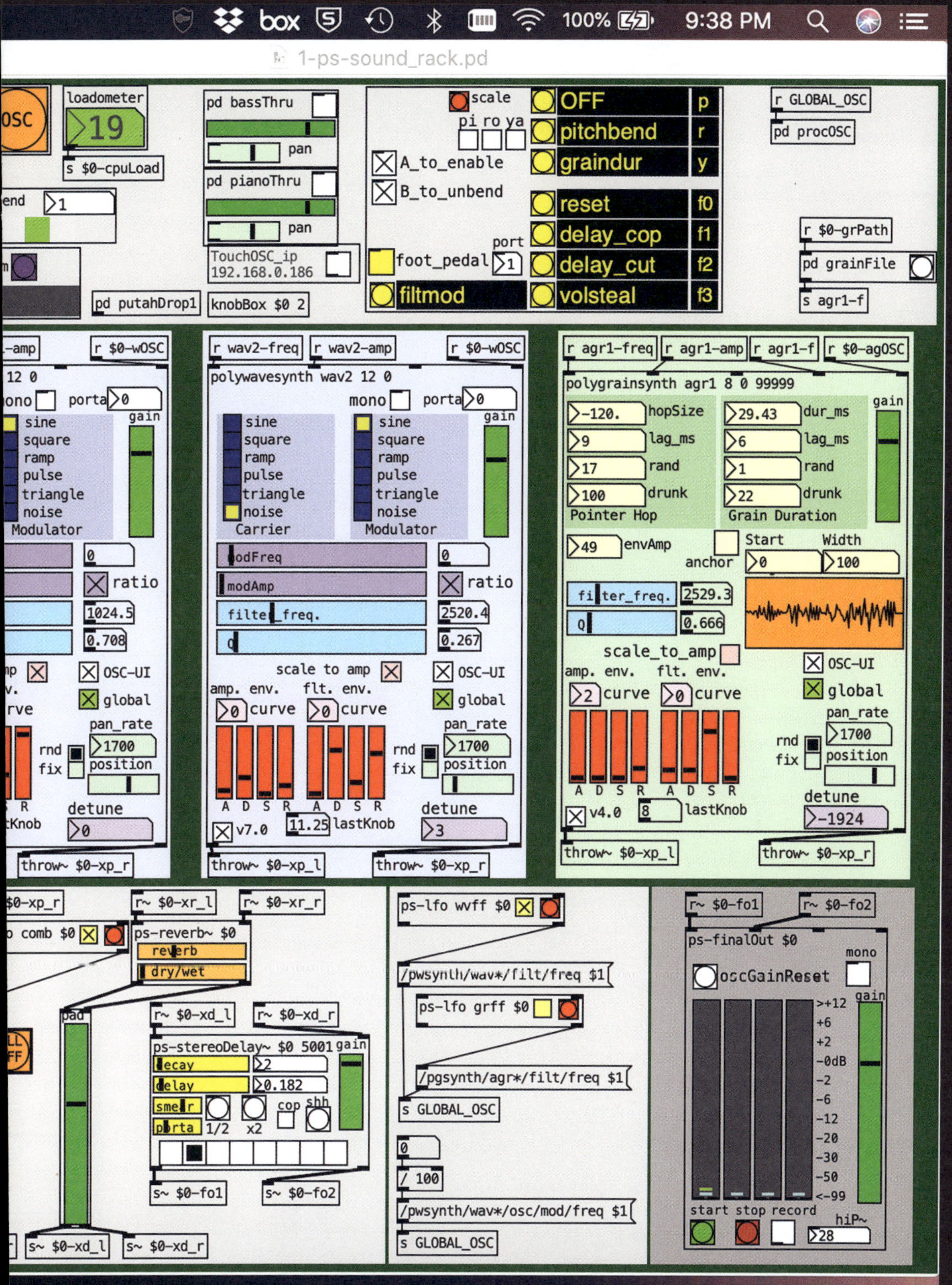

9:38 PM
1-ps-sound_rack.pd
loadometer
19
s $0-cpuLoad
pd bassThru
pan
pd pianoThru
pan
TouchOSC_ip
192.168.0.186
pd putahDrop1
knobBox $0 2
scale
pi ro ya
A_to_enable
B_to_unbend
foot_pedal
port
filtmod
OFF
pitchbend
graindur
reset
delay_cop
delay_cut
volsteal
p
r
y
f0
f1
f2
f3
r GLOBAL_OSC
pd procOSC
r $0-grPath
pd grainFile
s agr1-f
r $0-wOSC
r wav2-freq
r wav2-amp
polywavesynth wav2 12 0
mono
porta
gain
sine
square
ramp
pulse
triangle
noise
Carrier
Modulator
modFreq
modAmp
filter_freq.
ratio
1024.5
0.708
2520.4
0.267
scale to amp
amp. env.
flt. env.
curve
OSC-UI
global
pan_rate
1700
rnd
fix
position
detune
A D S R
v7.0
11.25
lastKnob
throw~ $0-xp_l
throw~ $0-xp_r
r agr1-freq
r agr1-amp
r agr1-f
r $0-agOSC
polygrainsynth agr1 8 0 99999
-120.
hopSize
lag_ms
rand
drunk
Pointer Hop
29.43
dur_ms
Grain Duration
envAmp
anchor
Start
Width
100
2529.3
0.666
scale_to_amp
v4.0
-1924
r~ $0-xr_l
r~ $0-xr_r
comb $0
ps-reverb~ $0
reverb
dry/wet
r~ $0-xd_l
r~ $0-xd_r
ps-stereoDelay~ $0 5001
decay
delay
smear
porta
0.182
1/2
x2
cop
shh
s~ $0-fo1
s~ $0-fo2
s~ $0-xd_l
s~ $0-xd_r
ps-lfo wvff $0
/pwsynth/wav*/filt/freq $1
ps-lfo grff $0
/pgsynth/agr*/filt/freq $1
s GLOBAL_OSC
/ 100
/pwsynth/wav*/osc/mod/freq $1
r~ $0-fo1
r~ $0-fo2
ps-finalOut $0
mono
oscGainReset
>+12
+6
+2
-0dB
-2
-6
-12
-20
-30
-50
<-99
start stop record
hiP~
28

Synertek Systems Corp.
SYM
MODEL 1
SYNERTEK SYSTEMS SYM
MC6850P
MC68B50P
UM6522A
LD 1
LD 2
SAV 1
SAV 2
RST
ON
GO
REG
MEM
OFF
CR
SHIFT
C D E F
8 9 A B
4 5 6 7
0 1 2 3
0027